Praise for Ian Stewart's *In Pursuit of the Unknown*

"Seemingly basic equations have enabled us to predict eclipses, engineer earthquake-proof buildings, and invent the refrigerator. In this lively volume, mathematician Ian Stewart delves into 17 equations that shape our daily existence, including those dreamed up by the likes of Einstein, Newton, and Erwin Schrödinger." —*Discover*

"*In Pursuit of the Unknown* is a really fun read. Ian Stewart takes 17 equations and instead of providing a tried-and-true linear narrative provides a more interesting, quirky, nonlinear one, going from formula to back-history to a few relevant digressions, to the formula's impact on the world with a few more digressions along the way. . . . *In Pursuit of the Unknown* is more about the wonders of scientific discovery than it is about anything else, and Ian Stewart is a genius in the way he conveys his excitement and sense of wonder. . . . He has that valuable grasp of not only what it takes to make equations interesting, but also to make science *cool*."
—*New York Journal of Books*

"Stewart's well-deserved reputation as a popularizer of mathematics and science is brilliantly upheld in this intriguing book. As the title suggests, each chapter explores an equation, but the power of the presentation comes with his clarifications regarding what each equation addresses, why it is important, and what it has led to." —*Choice*

"An interesting and highly entertaining book." —*Physics Today*

"[An] engaging work. . . . An entertaining journey through the development of theoretical mathematics, as well as an informative look at applied science." —*Book News*

"If you were a mathlete or have even the slightest interest in figures, Ian Stewart can bring the subject alive for you." —*History Wire*

"From how the Pythagorean theorem, which linked geometry and algebra, laid the groundwork of the best current theories of space, time, and gravity to how the Navier-Stokes equation applies to modeling climate changes, Stewart delivers a scientist's gift in a storyteller's package to reveal how these seemingly esoteric equations are really the foundation for nearly everything we know and use today. . . . Far from being a mere math primer or trivia aid, *In Pursuit of the Unknown* is an essential piece of modern literacy, wrapped in an articulate argument for why this kind of knowledge *should* be precisely that." —*Brain Pickings*

"Ian Stewart has penned a book for both sides of the divide with *In Pursuit of the Unknown,* ably bridging the gap between those daunted and undaunted by the language of equations. . . . By making some tough concepts truly accessible, *In Pursuit of the Unknown* will hopefully open a few more minds to possibility." —*Tulsa Book Review*

"Stewart's expertise and his well-developed style (enhanced by a nice sense of humor) make for enjoyable reading. . . . A worthwhile and entertaining book, accessible to all readers. Recommended for anyone interested in the influence of mathematics on the development of science and on the emergence of our current technology-driven society."
—*Library Journal*

"Stewart provides clear, cogent explanations of how the equations work without burdening the reader with cumbersome derivations. . . . He gives a fascinating explanation of how Newton's laws, when extended to three-body problems, are still used by NASA to calculate the best route from Earth to Mars and have laid the basis for chaos theory. Throughout, Stewart's style is felicitous."
—*Kirkus Reviews*

"Stewart shares his enthusiasm as well as his knowledge in this tour of ground-breaking equations and the research they supported. . . . Stewart assembles an entertaining and illuminating collection of curious facts and histories suitable for random dipping-in or reading straight through."
—*Publishers Weekly*

IN PURSUIT OF THE UNKNOWN

By the same author

Concepts of Modern Mathematics
Game, Set, and Math
The Problems of Mathematics
Does God Play Dice?
Another Fine Math You've Got Me into
Fearful Symmetry (with Martin Golubitsky)
Nature's Numbers
From Here to Infinity
The Magical Maze
Life's Other Secret
Flatterland
What Shape Is a Snowflake?
The Annotated Flatland
Math Hysteria
The Mayor of Uglyville's Dilemma
Letters to a Young Mathematician
Why Beauty Is Truth
How to Cut a Cake
Taming the Infinite/The Story of Mathematics
Professor Stewart's Cabinet of Mathematical Curiosities
Professor Stewart's Hoard of Mathematical Treasures
Cows in the Maze
Mathematics of Life
The Great Mathematical Problems

with Terry Pratchett and Jack Cohen
The Science of Discworld
The Science of Discworld II: the Globe
The Science of Discworld III: Darwin's Watch

with Jack Cohen
The Collapse of Chaos
Figments of Reality
Evolving the Alien/What Does a Martian Look Like?
Wheelers (science fiction)
Heaven (science fiction)

IN PURSUIT OF THE
UNKNOWN

17 Equations
That Changed the World

IAN STEWART

BASIC BOOKS
A Member of the Perseus Books Group
New York

Hardcover first published in 2012 in the United States by Basic Books,
A Member of the Perseus Books Group
Paperback first published in 2013 by Basic Books

Published in Great Britain in 2012 by Profile Books

Books published by Basic Books are available at special discounts for bulk purchases in the United States by corporations, institutions, and other organizations.
For more information, please contact the Special Markets Department at the
Perseus Books Group, 2300 Chestnut Street, Suite 200, Philadelphia, PA 19103, or
call (800) 810-4145, ext. 5000, or
e-mail special.markets@perseusbooks.com.

A CIP catalog record for this book is available from the Library of Congress.

LCCN: 2011944850
Hardcover ISBN: 978-0-465-02973-0
Paperback ISBN: 978-0-465-08598-9
E-book ISBN: 978-0-465-02974-7

10 9 8 7 6

Contents

To avoide the tediouse repetition of these woordes: is equalle to: I will sette as I doe often in woorke use, a paire of paralleles, or gemowe lines of one lengthe: ═════, bicause noe .2. thynges, can be moare equalle.

Robert Recorde, *The Whetstone of Witte*, 1557

Why Equations?

Equations are the lifeblood of mathematics, science, and technology. Without them, our world would not exist in its present form. However, equations have a reputation for being scary: Stephen Hawking's publishers told him that every equation would halve the sales of *A Brief History of Time*, but then they ignored their own advice and allowed him to include $E = mc^2$ when cutting it out would allegedly have sold another 10 million copies. I'm on Hawking's side. Equations are too important to be hidden away. But his publishers had a point too: equations are formal and austere, they look complicated, and even those of us who love equations can be put off if we are bombarded with them.

In this book, I have an excuse. Since it's *about* equations, I can no more avoid including them than I could write a book about mountaineering without using the word 'mountain'. I want to convince you that equations have played a vital part in creating today's world, from mapmaking to satnav, from music to television, from discovering America to exploring the moons of Jupiter. Fortunately, you don't need to be a rocket scientist to appreciate the poetry and beauty of a good, significant equation.

There are two kinds of equations in mathematics, which on the surface look very similar. One kind presents relations between various mathematical quantities: the task is to prove the equation is true. The other kind provides information about an unknown quantity, and the mathematician's task is to *solve* it – to make the unknown known. The distinction is not clear-cut, because sometimes the same equation can be used in both ways, but it's a useful guideline. You will find both kinds here.

Equations in pure mathematics are generally of the first kind: they reveal deep and beautiful patterns and regularities. They are valid because, given our basic assumptions about the logical structure of mathematics, there is no alternative. Pythagoras's theorem, which is an equation expressed in the language of geometry, is an example. If you accept Euclid's basic assumptions about geometry, then Pythagoras's theorem is *true*.

Equations in applied mathematics and mathematical physics are usually of the second kind. They encode information about the real world; they express properties of the universe that could in principle have been very different. Newton's law of gravity is a good example. It tells us how the attractive force between two bodies depends on their masses, and how far apart they are. Solving the resulting equations tells us how the planets orbit the Sun, or how to design a trajectory for a space probe. But Newton's law isn't a mathematical theorem; it's true for physical reasons, it fits observations. The law of gravity might have been different. Indeed, it *is* different: Einstein's general theory of relativity improves on Newton by fitting some observations better, while not messing up those where we already know Newton's law does a good job.

The course of human history has been redirected, time and time again, by an equation. Equations have hidden powers. They reveal the innermost secrets of nature. This is not the traditional way for historians to organise the rise and fall of civilisations. Kings and queens and wars and natural disasters abound in the history books, but equations are thin on the ground. This is unfair. In Victorian times, Michael Faraday was demonstrating connections between magnetism and electricity to audiences at the Royal Institution in London. Allegedly, Prime Minister William Gladstone asked whether anything of practical consequence would come from it. It is said (on the basis of very little actual evidence, but why ruin a nice story?) that Faraday replied: 'Yes, sir. One day you will tax it.' If he did say that, he was right. James Clerk Maxwell transformed early experimental observations and empirical laws about magnetism and electricity into a system of equations for electromagnetism. Among the many consequences were radio, radar, and television.

An equation derives its power from a simple source. It tells us that two calculations, which appear different, have the same answer. The key symbol is the equals sign, $=$. The origins of most mathematical symbols are either lost in the mists of antiquity, or are so recent that there is no doubt where they came from. The equals sign is unusual because it dates back more than 450 years, yet we not only know who invented it, we even know *why*. The inventor was Robert Recorde, in 1557, in *The Whetstone of Witte*. He used two parallel lines (he used an obsolete word *gemowe*, meaning 'twin') to avoid tedious repetition of the words 'is equal to'. He chose that symbol because 'no two things can be more equal'. Recorde chose well. His symbol has remained in use for 450 years.

The power of equations lies in the philosophically difficult correspondence between mathematics, a collective creation of human minds, and an external physical reality. Equations model deep patterns in the outside world. By learning to value equations, and to read the stories they tell, we can uncover vital features of the world around us. In principle, there might be other ways to achieve the same result. Many people prefer words to symbols; language, too, gives us power over our surroundings. But the verdict of science and technology is that words are too imprecise, and too limited, to provide an effective route to the deeper aspects of reality. They are too coloured by human-level assumptions. Words alone can't provide the essential insights.

Equations can. They have been a prime mover in human civilisation for thousands of years. Throughout history, equations have been pulling the strings of society. Tucked away behind the scenes, to be sure – but the influence was there, whether it was noticed or not. This is the story of the ascent of humanity, told through 17 equations.

1 The squaw on the hippopotamus
Pythagoras's Theorem

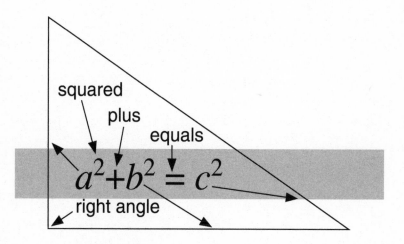

squared

plus

equals

$$a^2 + b^2 = c^2$$

right angle

What does it tell us?

How the three sides of a right-angled triangle are related.

Why is that important?

It provides a vital link between geometry and algebra, allowing us to calculate distances in terms of coordinates. It also inspired trigonometry.

What did it lead to?

Surveying, navigation, and more recently special and general relativity – the best current theories of space, time, and gravity.

A sk any school student to name a famous mathematician, and, assuming they can think of one, more often than not they will opt for Pythagoras. If not, Archimedes might spring to mind. Even the illustrious Isaac Newton has to play third fiddle to these two superstars of the ancient world. Archimedes was an intellectual giant, and Pythagoras probably wasn't, but he deserves more credit than he is often given. Not for what he achieved, but for what he set in motion.

Pythagoras was born on the Greek island of Samos, in the eastern Aegean, around 570 BC. He was a philosopher and a geometer. What little we know about his life comes from much later writers and its historical accuracy is questionable, but the key events are probably correct. Around 530 BC he moved to Croton, a Greek colony in what is now Italy. There he founded a philosophico-religious cult, the Pythagoreans, who believed that the universe is based on number. Their founder's present-day fame rests on the theorem that bears his name. It has been taught for more than 2000 years, and has entered popular culture. The 1958 movie *Merry Andrew*, starring Danny Kaye, includes a song whose lyrics begin:

> *The square on the hypotenuse*
> *of a right triangle*
> *is equal to*
> *the sum of the squares*
> *on the two adjacent sides.*

The song goes on with some *double entendre* about not letting your participle dangle, and associates Einstein, Newton, and the Wright brothers with the famous theorem. The first two exclaim 'Eureka!'; no, that was Archimedes. You will deduce that the lyrics are not hot on historical accuracy, but that's Hollywood for you. However, in Chapter 13 we will see that the lyricist (Johnny Mercer) was spot on with Einstein, probably more so than he realised.

Pythagoras's theorem appears in a well-known joke, with terrible puns about the squaw on the hippopotamus. The joke can be found all over the internet, but it's much harder to discover where it came from.[1] There are Pythagoras cartoons, T-shirts, and a Greek stamp, Figure 1.

Fig 1 Greek stamp showing Pythagoras's theorem.

All this fuss notwithstanding, we have no idea whether Pythagoras actually *proved* his theorem. In fact, we don't know whether it was his theorem at all. It could well have been discovered by one of Pythagoras's minions, or some Babylonian or Sumerian scribe. But Pythagoras got the credit, and his name stuck. Whatever its origins, the theorem and its consequences have had a gigantic impact on human history. They literally opened up our world.

The Greeks did not express Pythagoras's theorem as an equation in the modern symbolic sense. That came later with the development of algebra. In ancient times, the theorem was expressed verbally and geometrically. It attained its most polished form, and its first recorded proof, in the writings of Euclid of Alexandria. Around 250 BC Euclid became the first modern mathematician when he wrote his famous *Elements*, the most influential mathematical textbook ever. Euclid turned geometry into logic by making his basic assumptions explicit and invoking them to give systematic proofs for all of his theorems. He built a conceptual tower whose foundations were points, lines, and circles, and whose pinnacle was the existence of precisely five regular solids.

One of the jewels in Euclid's crown was what we now call Pythagoras's theorem: Proposition 47 of Book I of the *Elements*. In the famous translation by Sir Thomas Heath this proposition reads: 'In right-angled triangles the square on the side subtending the right angle is equal to the squares on the sides containing the right angle.'

No hippopotamus, then. No hypotenuse. Not even an explicit 'sum' or 'add'. Just that funny word 'subtend', which basically means 'be opposite to'. However, Pythagoras's theorem clearly expresses an equation, because it contains that vital word: *equal*.

For the purposes of higher mathematics, the Greeks worked with lines and areas instead of numbers. So Pythagoras and his Greek successors would decode the theorem as an equality of areas: 'The area of a square constructed using the longest side of a right-angled triangle is the sum of the areas of the squares formed from the other two sides.' The longest side is the famous hypotenuse, which means 'to stretch under', which it does if you draw the diagram in the appropriate orientation, as in Figure 2 (left).

Within a mere 2000 years, Pythagoras's theorem had been recast as the algebraic equation

$$a^2 + b^2 = c^2$$

where c is the length of the hypotenuse, a and b are the lengths of the other two sides, and the small raised 2 means 'square'. Algebraically, the square of any number is that number multiplied by itself, and we all know that the area of any square is the square of the length of its side. So Pythagoras's equation, as I shall rename it, says the same thing that Euclid said – except for various psychological baggage to do with how the ancients thought about basic mathematical concepts like numbers and areas, which I won't go into.

Pythagoras's equation has many uses and implications. Most directly, it lets you calculate the length of the hypotenuse, given the other two sides. For instance, suppose that $a = 3$ and $b = 4$. Then $c^2 = a^2 + b^2 = 3^2 + 4^2 = 9 + 16 = 25$. Therefore $c = 5$. This is the famous 3–4–5 triangle, ubiquitous in school mathematics, and the simplest example of a Pythagorean triple: a list of three whole numbers that satisfies Pythagoras's equation. The next simplest, other than scaled versions such as 6–8–10, is the 5–12–13 triangle. There are infinitely many such triples, and the Greeks knew how to construct them all. They still retain some interest in number theory, and even in the last decade new features have been discovered.

Instead of using a and b to work out c, you can proceed indirectly, and solve the equation to obtain a provided that you know b and c. You can also answer more subtle questions, as we will shortly see.

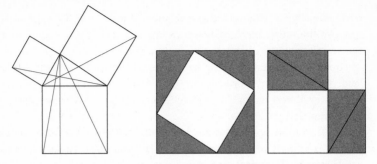

Fig 2 *Left*: Construction lines for Euclid's proof of Pythagoras. *Middle* and *right*: Alternative proof of the theorem. The outer squares have equal areas, and the shaded triangles all have equal areas. Therefore the tilted white square has the same area as the other two white squares combined.

Why is the theorem true? Euclid's proof is quite complicated, and it involves drawing five extra lines on the diagram, Figure 2 (left), and appealing to several previously proved theorems. Victorian schoolboys (few girls did geometry in those days) referred to it irreverently as Pythagoras's pants. A straightforward and intuitive proof, though not the most elegant, uses four copies of the triangle to relate two solutions of the same mathematical jigsaw puzzle, Figure 2 (right). The picture is compelling, but filling in the logical details requires some thought. For instance: how do we know that the tilted white region in the middle picture is a square?

There is tantalising evidence that Pythagoras's theorem was known long before Pythagoras. A Babylonian clay tablet[2] in the British Museum contains, in cuneiform script, a mathematical problem and answer that can be paraphrased as:

> 4 is the length and 5 the diagonal. What is the breadth?
> 4 times 4 is 16.
> 5 times 5 is 25.
> Take 16 from 25 to obtain 9.
> What times what must I take to get 9?
> 3 times 3 is 9.
> Therefore 3 is the breadth.

So the Babylonians certainly knew about the 3–4–5 triangle, a thousand years before Pythagoras.

Another tablet, YBC 7289 from the Babylonian collection of Yale University, is shown in Figure 3 (*left*). It shows a diagram of a square of side 30, whose diagonal is marked with two lists of numbers: 1, 24, 51, 10 and 42, 25, 35. The Babylonians employed base-60 notation for numbers, so the first list actually refers to $1 + 24/60 + 51/60^2 + 10/60^3$, which in decimals is 1.4142129. The square root of 2 is 1.4142135. The second list is 30 times this. So the Babylonians knew that the diagonal of a square is its side multiplied by the square root of 2. Since $1^2 + 1^2 = 2 = (\sqrt{2})^2$, this too is an instance of Pythagoras's theorem.

Fig 3 *Left*: YBC 7289. *Right*: Plimpton 322.

Even more remarkable, though more enigmatic, is the tablet Plimpton 322 from George Arthur Plimpton's collection at Columbia University, Figure 3 (right). It is a table of numbers, with four columns and 15 rows. The final column just lists the row number, from 1 to 15. In 1945 historians of science Otto Neugebauer and Abraham Sachs[3] noticed that in each row, the square of the number (say c) in the third column, minus the square of the number (say b) in the second column, is itself a square (say a). It follows that $a^2 + b^2 = c^2$, so the table appears to record Pythagorean triples. At least, this is the case provided four apparent errors are corrected. However, it is not absolutely certain that Plimpton 322 has anything to do with Pythagorean triples, and even if it does, it might just have been a convenient list of triangles whose areas were easy to calculate. These could then be assembled to yield good approximations to other triangles and other shapes, perhaps for land measurement.

Another iconic ancient civilisation is that of Egypt. There is some evidence that Pythagoras may have visited Egypt as a young man, and some have conjectured that this is where he learned his theorem. The surviving records of Egyptian mathematics offer scant support for this idea, but they are few and specialised. It is often stated, typically in the context of pyramids, that the Egyptians laid out right angles using a 3–4–5 triangle, formed from a length of string with knots at 12 equal intervals, and that archaeologists have found strings of that kind. However, neither claim makes much sense. Such a technique would not be very reliable, because strings can stretch and the knots would have to be very accurately spaced. The precision with which the pyramids at Giza are built is superior to anything that could be achieved with such a string. Far more practical tools, similar to a carpenter's set square, have been found. Egyptologists specialising in ancient Egyptian mathematics know of no records of string being employed to form a 3–4–5 triangle, and no examples of such strings exist. So this story, charming though it may be, is almost certainly a myth.

If Pythagoras could be transplanted into today's world he would notice many differences. In his day, medical knowledge was rudimentary, lighting came from candles and burning torches, and the fastest forms of communication were a messenger on horseback or a lighted beacon on a hilltop. The known world encompassed much of Europe, Asia, and Africa – but not the Americas, Australia, the Arctic, or the Antarctic. Many cultures considered the world to be flat: a circular disc or even a square aligned with the four cardinal points. Despite the discoveries of classical Greece this belief was still widespread in medieval times, in the form of *orbis terrae* maps, Figure 4.

Who first realised the world is round? According to Diogenes Laertius, a third-century Greek biographer, it was Pythagoras. In his book *Lives and Opinions of Eminent Philosophers*, a collection of sayings and biographical notes that is one of our main historical sources for the private lives of the philosophers of ancient Greece, he wrote: 'Pythagoras was the first who called the Earth round, though Theophrastus attributes this to Parmenides and Zeno to Hesiod.' The ancient Greeks often claimed that major discoveries had been made by their famous forebears, irrespective of historical fact, so we can't take the statement at face value, but it is not in dispute that from the fifth century BC all reputable Greek philosophers and mathematicians considered the Earth to be round. The idea does seem to have originated around the time of Pythagoras, and it might have come from one of his followers. Or it might have been common currency, based

Fig 4 Map of the world made around 1100 by the Moroccan cartographer al-Idrisi for King Roger of Sicily.

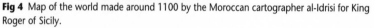

on evidence such as the round shadow of the Earth on the Moon during an eclipse, or the analogy with an obviously round Moon.

Even for the Greeks, though, the Earth was the centre of the universe and everything else revolved around it. Navigation was carried out by dead reckoning: looking at the stars and following the coastline. Pythagoras's equation changed all that. It set humanity on the path to today's understanding of the geography of our planet and its place in the Solar System. It was a vital first step towards the geometric techniques needed for mapmaking, navigation, and surveying. It also provided the key to a vitally important relation between geometry and algebra. This line of development leads from ancient times right through to general relativity and modern cosmology, see Chapter 13. Pythagoras's equation opened up entirely new directions for human exploration, both metaphorically and literally. It revealed the shape of our world and its place in the universe.

Many of the triangles encountered in real life are not right-angled, so the equation's direct applications may seem limited. However, any triangle can be cut into two right-angled ones as in Figure 6 (page 11), and any polygonal shape can be cut into triangles. So right-angled triangles are the key: they prove that there is a useful relation between the shape of a triangle and the lengths of its sides. The subject that developed from this insight is trigonometry: 'triangle measurement'.

The right-angled triangle is fundamental to trigonometry, and in particular it determines the basic trigonometric functions: sine, cosine, and tangent. The names are Arabic in origin, and the history of these functions and their many predecessors shows the complicated route by which today's version of the topic arose. I'll cut to the chase and explain the eventual outcome. A right-angled triangle has, of course, a right angle, but its other two angles are arbitrary, apart from adding to $90°$. Associated with any angle are three functions, that is, rules for calculating an associated number. For the angle marked A in Figure 5, using the traditional a, b, c for the three sides, we define the sine (sin), cosine (cos), and tangent (tan) like this:

$$\sin A = a/c \quad \cos A = b/c \quad \tan A = a/b$$

These quantities depend only on the angle A, because all right-angled triangles with a given angle A are identical except for scale.

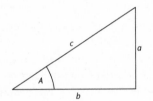

Fig 5 Trigonometry is based on a right-angle triangle.

In consequence, it is possible to draw up a table of the values of sin, cos, and tan, for a range of angles, and then use them to calculate features of right-angled triangles. A typical application, which goes back to ancient times, is to calculate the height of a tall column using only measurements made on the ground. Suppose that, from a distance of 100 metres, the angle to the top of the column is $22°$. Take $A = 22°$ in Figure 5, so that a is the height of the column. Then the definition of the tangent function tells us that

$$\tan 22° = a/100$$

so that

$$a = 100 \tan 22°.$$

Since tan $22°$ is 0.404, to three decimal places, we deduce that $a = 40.4$ metres.

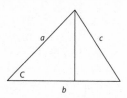

Fig 6 Splitting a triangle into two with right angles.

Once in possession of trigonometric functions, it is straightforward to extend Pythagoras's equation to triangles that do not have a right angle. Figure 6 shows a triangle with an angle C and sides a, b, c. Split the triangle into two right-angled ones as shown. Then two applications of Pythagoras and some algebra[4] prove that

$$a^2 + b^2 - 2ab \cos C = c^2$$

which is similar to Pythagoras's equation, except for the extra term $-2ab$ cos C. This 'cosine rule' does the same job as Pythagoras, relating c to a and b, but now we have to include information about the angle C.

The cosine rule is one of the mainstays of trigonometry. If we know two sides of a triangle and the angle between them, we can use it to calculate the third side. Other equations then tell us the remaining angles. All of these equations can ultimately be traced back to right-angled triangles.

Armed with trigonometric equations and suitable measuring apparatus, we can carry out surveys and make accurate maps. This is not a new idea. It appears in the Rhind Papyrus, a collection of ancient Egyptian mathematical techniques dating from 1650 BC. The Greek philosopher Thales used the geometry of triangles to estimate the heights of the Giza pyramids in about 600 BC. Hero of Alexandria described the same technique in 50 AD. Around 240 BC Greek mathematician, Eratosthenes, calculated the size of the Earth by observing the angle of the Sun at noon in two different places: Alexandria and Syene (now Aswan) in Egypt. A succession of Arabian scholars preserved and developed these methods, applying them in particular to astronomical measurements such as the size of the Earth.

Surveying began to take off in 1533 when the Dutch mapmaker Gemma Frisius explained how to use trigonometry to produce accurate maps, in *Libellus de Locorum Describendorum Ratione* ('Booklet Concerning a

Way of Describing Places'). Word of the method spread across Europe, reaching the ears of the Danish nobleman and astronomer Tycho Brahe. In 1579 Tycho used it to make an accurate map of Hven, the island where his observatory was located. By 1615 the Dutch mathematician Willebrord Snellius (Snel van Royen) had developed the method into essentially its modern form: *triangulation*. The area being surveyed is covered with a network of triangles. By measuring one initial length very carefully, and many angles, the locations of the corners of the triangle, and hence any interesting features within them, can be calculated. Snellius worked out the distance between two Dutch towns, Alkmaar and Bergen op Zoom, using a network of 33 triangles. He chose these towns because they lay on the same line of longitude and were exactly one degree of arc apart. Knowing the distance between them, he could work out the size of the Earth, which he published in his *Eratosthenes Batavus* ('The Dutch Eratosthenes') in 1617. His result is accurate to within 4%. He also modified the equations of trigonometry to reflect the spherical nature of the Earth's surface, an important step towards effective navigation.

Triangulation is an indirect method for calculating distances using angles. When surveying a stretch of land, be it a building site or a country, the main practical consideration is that it is much easier to measure angles than it is to measure distances. Triangulation lets us measure a few distances and lots of angles; then everything else follows from the trigonometric equations. The method begins by setting out one line between two points, called the baseline, and measuring its length directly to very high accuracy. Then we choose a prominent point in the landscape that is visible from both ends of the baseline, and measure the angle from the baseline to that point, at both ends of the baseline. Now we have a triangle, and we know one side of it and two angles, which fix its shape and size. We can then use trigonometry to work out the other two sides.

In effect, we now have two more baselines: the newly calculated sides of the triangle. From those, we can measure angles to other, more distant points. Continue this process to create a network of triangles that covers the area being surveyed. Within each triangle, observe the angles to all noteworthy features – church towers, crossroads, and so on. The same trigonometric trick pinpoints their precise locations. As a final twist, the accuracy of the entire survey can be checked by measuring one of the final sides directly.

By the late eighteenth century, triangulation was being employed routinely in surveys. The Ordnance Survey of Great Britain began in 1783, taking 70 years to complete the task. The Great Trigonometric Survey of

India, which among other things mapped the Himalayas and determined the height of Mount Everest, started in 1801. In the twenty-first century, most large-scale surveying is done using satellite photographs and GPS (the Global Positioning System). Explicit triangulation is no longer employed. But it is still there, behind the scenes, in the methods used to deduce locations from the satellite data.

Pythagoras's theorem was also vital to the invention of coordinate geometry. This is a way to represent geometric figures in terms of numbers, using a system of lines, known as axes, labelled with numbers. The most familiar version is known as Cartesian coordinates in the plane, in honour of the French mathematician and philosopher René Descartes, who was one of the great pioneers in this area – though not the first. Draw two lines: a horizontal one labelled x and a vertical one labelled y. These lines are known as axes (plural of axis), and they cross at a point called the origin. Mark points along these two axes according to their distance from the origin, like the markings on a ruler: positive numbers to the right and up, negative to the left and down. Now we can determine any point in the plane in terms of two numbers x and y, its coordinates, by connecting the point to the two axes as in Figure 7. The pair of numbers (x, y) completely specifies the location of the point.

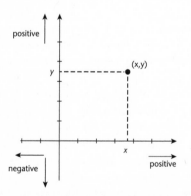

Fig 7 The two axes and the coordinates of a point.

The great mathematicians of seventeenth-century Europe realised that, in this context, a line or curve in the plane corresponds to the set of solutions (x, y) of some equation in x and y. For instance, $y = x$ determines a diagonal line sloping from lower left to top right, because (x, y) lies on that line if and only if $y = x$. In general, a linear equation – of the form $ax + by = c$ for constants a, b, c – corresponds to a straight line, and vice versa.

What equation corresponds to a circle? This is where Pythagoras's equation comes in. It implies that the distance r from the origin to the point (x, y) satisfies

$$r^2 = x^2 + y^2$$

and we can solve this for r to obtain

$$r = \sqrt{x^2 + y^2}$$

Since the set of all points that lie at distance r from the origin is a circle of radius r, whose centre is the origin, so the same equation defines a circle. More generally, the circle of radius r with centre at (a, b) corresponds to the equation

$$(x - a)^2 + (y - b)^2 = r^2$$

and the same equation determines the distance r between the two points (a, b) and (x, y). So Pythagoras's theorem tells us two vital things: which equations yield circles, and how to calculate distances from coordinates.

Pythagoras's theorem, then, is important in its own right, but it exerts even more influence through its generalisations. Here I will pursue just one strand of these later developments to bring out the connection with relativity, to which we return in Chapter 13.

The proof of Pythagoras's theorem in Euclid's *Elements* places the theorem firmly within the realm of Euclidean geometry. There was a time when that phrase could have been replaced by just 'geometry', because it was generally assumed that Euclid's geometry was the true geometry of physical space. It was obvious. Like most things assumed to be obvious, it turned out to be false.

Euclid derived all of his theorems from a small number of basic assumptions, which he classified as definitions, axioms, and common notions. His set-up was elegant, intuitive, and concise, with one glaring exception, his fifth axiom: 'If a straight line falling on two straight lines makes the interior angles on the same side less than two right angles, the two straight lines, if produced indefinitely, meet on that side on which are the angles less than the two right angles.' It's a bit of a mouthful: Figure 8 may help.

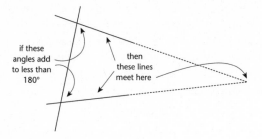

if these angles add to less than 180°

then these lines meet here

Fig 8 Euclid's parallel axiom.

For well over a thousand years, mathematicians tried to repair what they saw as a flaw. They weren't just looking for something simpler and more intuitive that would achieve the same end, although several of them found such things. They wanted to get rid of the awkward axiom altogether, by proving it. After several centuries, mathematicians finally realised that there were alternative 'non-Euclidean' geometries, implying that no such proof existed. These new geometries were just as logically consistent as Euclid's, and they obeyed all of his axioms except the parallel axiom. They could be interpreted as the geometry of geodesics – shortest paths – on curved surfaces, Figure 9. This focused attention on the meaning of curvature.

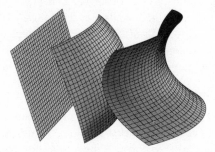

Fig 9 Curvature of a surface. *Left*: zero curvature. *Middle*: positive curvature. *Right*: negative curvature.

The plane of Euclid is flat, curvature zero. A sphere has the same curvature everywhere, and it is positive: near any point it looks like a dome. (As a technical fine point: great circles meet in two points, not one as Euclid's axioms require, so spherical geometry is modified by identifying antipodal points on the sphere – considering them to be identical. The surface becomes a so-called projective plane and the geometry is called elliptic.) A surface of constant negative curvature also exists: near any point, it looks like a saddle. This surface is called the hyperbolic plane, and it can be represented in several entirely prosaic ways. Perhaps the simplest is to consider it as the interior of a circular disc, and to define 'line' as an arc of a circle meeting the edge of the disc at right angles (Figure 10).

Fig 10 Disc model of the hyperbolic plane. All three lines through *P* fail to meet line *L*.

It might seem that, while plane geometry might be non-Euclidean, this must be impossible for the geometry of space. You can bend a surface by pushing it into a third dimension, but you can't bend *space* because there's no room for an extra dimension along which to push it. However, this is a rather naive view. For example, we can model three-dimensional hyperbolic space using the interior of a sphere. Lines are modelled as arcs of circles that meet the boundary at right angles, and planes are modelled as parts of spheres that meet the boundary at right angles. This geometry is three-dimensional, satisfies all of Euclid's axioms except the Fifth, and in a sense that can be pinned down it defines a curved three-dimensional space. But it's not curved round anything, or in any new direction.

It's just curved.

With all these new geometries available, a new point of view began to occupy centre stage – but as physics, not mathematics. Since space doesn't *have* to be Euclidean, what shape *is* it? Scientists realised that they didn't actually know. In 1813, Gauss, knowing that in a curved space the angles of a triangle do not add to 180°, measured the angles of a triangle formed by three mountains – the Brocken, the Hohehagen, and the Inselberg. He obtained a sum 15 seconds of arc greater than 180°. If correct, this indicated that space (in that region, at least) was positively curved. But you'd need a far larger triangle, and far more accurate measurements, to eliminate observational errors. So Gauss's observations were inconclusive. Space might be Euclidean, and then again, it might not be.

My remark that three-dimensional hyperbolic space is 'just curved' depends on a new point of view about curvature, which also goes back to Gauss. The sphere has constant positive curvature, and the hyperbolic plane has constant negative curvature. But the curvature of a surface doesn't have to be constant. It might be sharply curved in some places, less sharply curved in others. Indeed, it might be positive in some regions but negative in others. The curvature could vary continuously from place to place. If the surface looks like a dog's bone, then the blobs at the ends are positively curved but the part that joins them is negatively curved.

Gauss searched for a formula to characterise the curvature of a surface at any point. When he eventually found it, and published it in his *Disquisitiones Generales Circa Superficies Curva* ('General Research on Curved Surfaces') of 1828, he named it the 'remarkable theorem'. What was so remarkable? Gauss had started from the naive view of curvature: embed the surface in three-dimensional space and calculate how bent it is. But the answer told him that this surrounding space didn't matter. It didn't enter into the formula. He wrote: 'The formula ... leads itself to the remarkable theorem: If a curved surface is developed upon any other surface whatever, the measure of curvature in each point remains unchanged.' By 'developed' he meant 'wrapped round'.

Take a flat sheet of paper, zero curvature. Now wrap it round a bottle. If the bottle is cylindrical the paper fits perfectly, without being folded, stretched, or torn. It is bent as far as visual appearance goes, but it's a trivial kind of bending, because it hasn't changed geometry on the paper in any way. It's just changed how the paper relates to the surrounding space. Draw a right-angled triangle on the flat paper, measure its sides, check Pythagoras. Now wrap the diagram round a bottle. The lengths of sides, *measured along the paper*, don't change. Pythagoras is still true.

The surface of a sphere, however, has nonzero curvature. So it is not possible to wrap a sheet of paper so that it fits snugly against a sphere, without folding it, stretching it, or tearing it. Geometry on a sphere is intrinsically different from geometry on a plane. For example, the Earth's equator and the lines of longitude for $0°$ and $90°$ to its north determine a triangle that has three right angles and three equal sides (assuming the Earth to be a sphere). So Pythagoras's equation is false.

Today we call curvature in its intrinsic sense 'Gaussian curvature'. Gauss explained why it is important using a vivid analogy, still current. Imagine an ant confined to the surface. How can it work out whether the surface is curved? It can't step outside the surface to see whether it looks bent. But it can use Gauss's formula by making suitable measurements purely within the surface. We are in the same position as the ant when we try to figure out the true geometry of our space. We can't step outside it. Before we can emulate the ant by taking measurements, however, we need a formula for the curvature of a space of three dimensions. Gauss didn't have one. But one of his students, in a fit of recklessness, claimed that he did.

The student was Georg Bernhard Riemann, and he was trying to achieve what German universities call Habilitation, the next step after a PhD. In Riemann's day this meant that you could charge students a fee for your lectures. Then and now, gaining Habilitation requires presenting your research in a public lecture that is also an examination. The candidate offers several topics, and the examiner, which in Riemann's case was Gauss, chooses one. Riemann, a brilliant mathematical talent, listed several orthodox topics that he knew backwards, but in a rush of blood to the brain he also suggested 'On the hypotheses which lie at the foundation of geometry'. Gauss had long been interested in just that, and he naturally selected it for Riemann's examination.

Riemann instantly regretted offering something so challenging. He had a hearty dislike of public speaking, and he hadn't thought the mathematics through in detail. He just had some vague, though fascinating, ideas about curved space. In *any* number of dimensions. What Gauss had done for two dimensions, with his remarkable theorem, Riemann wanted to do in as many dimensions as you like. Now he had to perform, and fast. The lecture was looming. The pressure nearly gave him a nervous breakdown, and his day job helping Gauss's collaborator Wilhelm Weber with experiments in electricity didn't help. Well, maybe it did, because while Riemann was thinking about the relation between electrical and magnetic forces in the day job, he realised that force can be related to curvature. Working backwards, he could use the mathematics of forces to define curvature, as required for his examination.

In 1854 Riemann delivered his lecture, which was warmly received, and no wonder. He began by defining what he called a 'manifold', in the sense of many-foldedness. Formally, a 'manifold', is specified by a system of many coordinates, together with a formula for the distance between nearby points, now called a Riemannian metric. Informally, a manifold is a multidimensional space in all its glory. The climax of Riemann's lecture was a formula that generalised Gauss's remarkable theorem: it defined the curvature of the manifold solely in terms of its metric. And it is here that the tale comes full circle like the snake Orobouros and swallows its own tail, because the metric contains visible remnants of Pythagoras.

Suppose, for example, that the manifold has three dimensions. Let the coordinates of a point be (x, y, z), and let $(x+dx, y+dy, z+dz)$ be a nearby point, where the d means 'a little bit of'. If the space is Euclidean, with zero curvature, the distance ds between these two points satisfies the equation

$$ds^2 = dx^2 + dy^2 + dz^2$$

and this is just Pythagoras, restricted to points that are close together. If the space is curved, with variable curvature from point to point, the analogous formula, the metric, looks like this:

$$ds^2 = X\,dx^2 + Y\,dy^2 + Z\,dz^2 + 2U\,dx\,dy + 2V\,dx\,dz + 2W\,dy\,dz$$

Here X, Y, Z, U, V, W can depend on x, y and z. It may seem a bit of a mouthful, but like Pythagoras's equation it involves sums of squares (and closely related products of two quantities like $dx\,dy$) plus a few bells and whistles. The 2s occur because the formula can be packaged as a 3×3 table, or matrix:

$$\begin{bmatrix} X & U & V \\ U & Y & W \\ V & W & Z \end{bmatrix}$$

where X, Y, Z appear once, but U, V, W appear twice. The table is symmetric about its diagonal; in the language of differential geometry it is a symmetric tensor. Riemann's generalisation of Gauss's remarkable theorem is a formula for the curvature of the manifold, at any given point, in terms of this tensor. In the special case when Pythagoras applies, the curvature turns out to be zero. So the validity of Pythagoras's equation is a test for the absence of curvature.

Like Gauss's formula, Riemann's expression for curvature depends only on the manifold's metric. An ant confined to the manifold could observe the metric by measuring tiny triangles and computing the curvature. Curvature is an intrinsic property of a manifold, independent of any surrounding space. Indeed, the metric already determines the geometry, so no surrounding space is required. In particular, we human ants can ask what shape our vast and mysterious universe is, and hope to answer it by making observations that do not require us to step outside the universe. Which is just as well, because we can't.

Riemann found his formula by using forces to define geometry. Fifty years later, Einstein turned Riemann's idea on its head, using geometry to define the force of gravity in his general theory of relativity, and inspiring new ideas about the shape of the universe: see Chapter 13. It's an astonishing progression of events. Pythagoras's equation first came into being around 3500 years ago to measure a farmer's land. Its extension to triangles without right angles, and triangles on a sphere, allowed us to map our continents and measure our planet. And a remarkable generalisation lets us measure the shape of the universe. Big ideas have small beginnings.

2 Shortening the proceedings
Logarithms

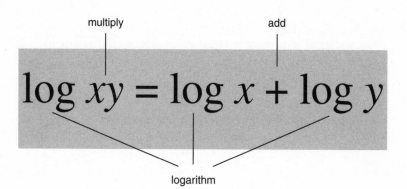

multiply   add

$$\log xy = \log x + \log y$$

logarithm

What does it tell us?

How to multiply numbers by adding related numbers instead.

Why is that important?

Addition is much simpler than multiplication.

What did it lead to?

Efficient methods for calculating astronomical phenomena such as eclipses and planetary orbits. Quick ways to perform scientific calculations. The engineers' faithful companion, the slide rule. Radioactive decay and the psychophysics of human perception.

Numbers originated in practical problems: recording property, such as animals or land, and financial transactions, such as taxation and keeping accounts. The earliest known number notation, aside from simple tallying marks like IIII, is found on the outside of clay envelopes. In 8000 BC Mesopotamian accountants kept records using small clay tokens of various shapes. The archaeologist Denise Schmandt-Besserat realised that each shape represented a basic commodity – a sphere for grain, an egg for a jar of oil, and so on. For security, the tokens were sealed in clay wrappings. But it was a nuisance to break a clay envelope open to find out how many tokens were inside, so the ancient accountants scratched symbols on the outside to show what was inside. Eventually they realised that once you had these symbols, you could scrap the tokens. The result was a series of written symbols for numbers – the origin of all later number symbols, and perhaps of writing too.

Along with numbers came arithmetic: methods for adding, subtracting, multiplying, and dividing numbers. Devices like the abacus were used to do the sums; then the results could be recorded in symbols. After a time, ways were found to use the symbols to perform the calculations without mechanical assistance, although the abacus is still widely used in many parts of the world, while electronic calculators have supplanted pen and paper calculations in most other countries.

Arithmetic proved essential in other ways, too, especially in astronomy and surveying. As the basic outlines of the physical sciences began to emerge, the fledgeling scientists needed to perform ever more elaborate calculations, by hand. Often this took up much of their time, sometimes months or years, getting in the way of more creative activities. Eventually it became essential to speed up the process. Innumerable mechanical devices were invented, but the most important breakthrough was a conceptual one: think first, calculate later. Using clever mathematics, you could make difficult calculations much easier.

The new mathematics quickly developed a life of its own, turning out to have deep theoretical implications as well as practical ones. Today, those early ideas have become an indispensable tool throughout science, reaching even into psychology and the humanities. They were widely used until the 1980s, when computers rendered them obsolete for practical purposes, but, despite that, their importance in mathematics and science has continued to grow.

The central idea is a mathematical technique called a logarithm. Its inventor was a Scottish laird, but it took a geometry professor with strong interests in navigation and astronomy to replace the laird's brilliant but flawed idea by a much better one.

In March 1615 Henry Briggs wrote a letter to James Ussher, recording a crucial event in the history of science:

> Napper, lord of Markinston, hath set my head and hands a work with his new and admirable logarithms. I hope to see him this summer, if it please God, for I never saw a book which pleased me better or made me more wonder.

Briggs was the first professor of geometry at Gresham College in London, and 'Napper, lord of Markinston' was John Napier, eighth laird of Merchiston, now part of the city of Edinburgh in Scotland. Napier seems to have been a bit of a mystic; he had strong theological interests, but they mostly centred on the book of Revelation. In his view, his most important work was *A Plaine Discovery of the Whole Revelation of St John*, which led him to predict that the world would end in either 1688 or 1700. He is thought to have engaged in both alchemy and necromancy, and his interests in the occult lent him a reputation as a magician. According to rumour, he carried a black spider in a small box everywhere he went, and possessed a 'familiar', or magical companion: a black cockerel. According to one of his descendants, Mark Napier, John employed his familiar to catch servants who were stealing. He locked the suspect in a room with the cockerel and instructed them to stroke it, telling them that his magical bird would unerringly detect the guilty. But Napier's mysticism had a rational core, which in this particular instance involved coating the cockerel with a thin layer of soot. An innocent servant would be confident enough to stroke the bird as instructed, and would get soot on their hands. A guilty one, fearing detection, would avoid stroking the bird. So, ironically, clean hands proved you were guilty.

Napier devoted much of his time to mathematics, especially methods for speeding up complicated arithmetical calculations. One invention,

Napier's bones, was a set of ten rods, marked with numbers, which simplified the process of long multiplication. Even better was the invention that made his reputation and created a scientific revolution: not his book on Revelation, as he had hoped, but his *Mirifici Logarithmorum Canonis Descriptio* ('Description of the Wonderful Canon of Logarithms') of 1614. The preface shows that Napier knew exactly what he had produced, and what it was good for.[1]

> Since nothing is more tedious, fellow mathematicians, in the practice of the mathematical arts, than the great delays suffered in the tedium of lengthy multiplications and divisions, the finding of ratios, and in the extraction of square and cube roots – and ... the many slippery errors that can arise: I had therefore been turning over in my mind, by what sure and expeditious art, I might be able to improve upon these said difficulties. In the end after much thought, finally I have found an amazing way of shortening the proceedings ... it is a pleasant task to set out the method for the public use of mathematicians.

The moment Briggs heard of logarithms, he was enchanted. Like many mathematicians of his era, he spent a lot of his time performing astronomical calculations. We know this because another letter from Briggs to Ussher, dated 1610, mentions calculating eclipses, and because Briggs had earlier published two books of numerical tables, one related to the North Pole and the other to navigation. All of these works had required vast quantities of complicated arithmetic and trigonometry. Napier's invention would save a great deal of tedious labour. But the more Briggs studied the book, the more convinced he became that although Napier's strategy was wonderful, he'd got his tactics wrong. Briggs came up with a simple but effective improvement, and made the long journey to Scotland. When they met, 'almost one quarter of an hour was spent, each beholding the other with admiration, before one word was spoken'.[2]

What was it that excited so much admiration? The vital observation, obvious to anyone learning arithmetic, was that adding numbers is relatively easy, but multiplying them is not. Multiplication requires many more arithmetical operations than addition. For example, adding two ten-digit numbers involves about ten simple steps, but multiplication requires 200. With modern computers, this issue is still important, but now it is tucked away behind the scenes in the algorithms used for multiplication.

But in Napier's day it all had to be done by hand. Wouldn't it be great if there were some mathematical trick that would convert those nasty multiplications into nice, quick addition sums? It sounds too good to be true, but Napier realised that it was possible. The trick was to work with powers of a fixed number.

In algebra, powers of an unknown x are indicated by a small raised number. That is, $xx = x^2$, $xxx = x^3$, $xxxx = x^4$, and so on, where as usual in algebra placing two letters next to each other means you should multiply them together. So, for instance, $10^4 = 10 \times 10 \times 10 \times 10 = 10{,}000$. You don't need to play around with such expressions for long before you discover an easy way to work out, say, $10^4 \times 10^3$. Just write down

$$10{,}000 \times 1{,}000 = (10 \times 10 \times 10 \times 10) \times (10 \times 10 \times 10)$$
$$= 10 \times 10 \times 10 \times 10 \times 10 \times 10 \times 10$$
$$= 10{,}000{,}000$$

The number of 0s in the answer is 7, which equals $4 + 3$. The first step in the calculation shows *why* it is $4 + 3$: we stick four 10s and three 10s next to each other. In short,

$$10^4 \times 10^3 = 10^{4+3} = 10^7$$

In the same way, whatever the value of x might be, if we multiply its ath power by its bth power, where a and b are whole numbers, then we get the $(a + b)$th power:

$$x^a \, x^b = x^{a+b}$$

This may seem an innocuous formula, but on the left it multiplies two quantities together, while on the right the main step is to add a and b, which is simpler.

Suppose you wanted to multiply, say, 2.67 by 3.51. By long multiplication you get 9.3717, which to two decimal places is 9.37. What if you try to use the previous formula? The trick lies in the choice of x. If we take x to be 1.001, then a bit of arithmetic reveals that

$$(1.001)^{983} = 2.67$$
$$(1.001)^{1256} = 3.51$$

correct to two decimal places. The formula then tells us that 2.87×3.41 is

$$(1.001)^{983 + 1256} = (1.001)^{2239}$$

which, to two decimal places, is 9.37.

The core of the calculation is an easy addition: $983 + 1256 = 2239$. However, if you try to check my arithmetic you will quickly realise that if anything I've made the problem harder, not easier. To work out $(1.001)^{983}$ you have to multiply 1.001 by itself 983 times. And to discover that 983 is the right power to use, you have to do even more work. So at first sight this seems like a pretty useless idea.

Napier's great insight was that this objection is wrong. But to overcome it, some hardy soul has to calculate lots of powers of 1.001, starting with $(1.001)^2$ and going up to something like $(1.001)^{10,000}$. Then they can publish a table of all these powers. After that, most of the work has been done. You just have to run your fingers down the successive powers until you see 2.67 next to 983; you similarly locate 3.51 next to 1256. Then you add those two numbers to get 2239. The corresponding row of the table tells you that this power of 1.001 is 9.37. Job done.

Really accurate results require powers of something a lot closer to 1, such as 1.000001. This makes the table far bigger, with a million or so powers. Doing the calculations for that table is a huge enterprise. *But it has to be done only once.* If some self-sacrificing benefactor makes the effort up front, succeeding generations will be saved a gigantic amount of arithmetic.

In the context of this example, we can say that the powers 983 and 1256 are the *logarithms* of the numbers 2.67 and 3.51 that we wish to multiply. Similarly 2239 is the logarithm of their product 9.38. Writing log as an abbreviation, what we have done amounts to the equation

$$\log ab = \log a + \log b$$

which is valid for any numbers a and b. The rather arbitrary choice of 1.001 is called the *base*. If we use a different base, the logarithms that we calculate are also different, but for any fixed base everything works the same way.

This is what Napier should have done. But for reasons that we can only guess at, he did something slightly different. Briggs, approaching the technique from a fresh perspective, spotted two ways to improve on Napier's idea.

When Napier started thinking about powers of numbers, in the late sixteenth century, the idea of reducing multiplication to addition was already circulating among mathematicians. A rather complicated method known as 'prosthapheiresis', based on a formula involving trigonometric functions, was in use in Denmark.[3] Napier, intrigued, was smart enough to realise that powers of a fixed number could do the same job more simply. The necessary tables didn't exist – but that was easily remedied. Some public-spirited soul must carry out the work. Napier volunteered himself for the task, but he made a strategic error. Instead of using a base that was slightly bigger than 1, he used a base slightly smaller than 1. In consequence, the sequence of powers started out with big numbers, which got successively smaller. This made the calculations slightly more clumsy.

Briggs spotted this problem, and saw how to deal with it: use a base slightly larger than 1. He also spotted a subtler problem, and dealt with that as well. If Napier's method were modified to work with powers of something like 1.0000000001, there would be no straightforward relation between the logarithms of, say, 12.3456 and 1.23456. So it wasn't entirely clear when the table could *stop*. The source of the problem was the value of log 10, because

$$\log 10x = \log 10 + \log x$$

Unfortunately log 10 was messy: with the base 1.0000000001 the logarithm of 10 was 23,025,850,929. Briggs thought it would be much nicer if the base could be chosen so that log 10 = 1. Then log $10x = 1 + \log x$, so that whatever log 1.23456 might be, you just had to add 1 to it to get log 12.3456. Now tables of logarithms need only run from 1 to 10. If bigger numbers turned up, you just added the appropriate whole number.

To make log 10 = 1, you do what Napier did, using a base of 1.0000000001, but then you divide every logarithm by that curious number 23,025,850,929. The resulting table consists of logarithms to base 10, which I'll write as $\log_{10} x$. They satisfy

$$\log_{10} xy = \log_{10} x + \log_{10} y$$

as before, but also

$$\log_{10} 10x = \log_{10} x + 1$$

Within two years Napier was dead, so Briggs started work on a table of base-10 logarithms. In 1617 he published *Logarithmorum Chilias Prima* ('Logarithms of the First Chiliad'), the logarithms of the integers from 1 to 1000 accurate to 14 decimal places. In 1624 he followed it up with *Arithmetic Logarithmica* ('Arithmetic of Logarithms'), a table of base-10 logarithms of numbers from 1 to 20,000 and from 90,000 to 100,000, to the same accuracy. Others rapidly followed Briggs's lead, filling in the large gap and developing auxiliary tables such as logarithms of trigonometric functions like log sin x.

The same ideas that inspired logarithms allow us to define powers x^a of a positive variable x for values of a that are not positive whole numbers. All we have to do is insist that our definitions must be consistent with the equation $x^a x^b = x^{a+b}$, and follow our noses. To avoid nasty complications, it is best to assume x is positive, and to define x^a so that this is also positive. (For negative x, it's best to introduce complex numbers, as in Chapter 5.)

For example, what is x^0? Bearing in mind that $x^1 = x$, the formula says that x^0 must satisfy $x^0 x = x^{0+1} = x$. Dividing by x we find that $x^0 = 1$. Now what about x^{-1}? Well, the formula says that $x^{-1} x = x^{-1+1} = x^0 = 1$. Dividing by x, we get $x^{-1} = 1/x$. Similarly $x^{-2} = 1/x^2$, $x^{-3} = 1/x^3$, and so on.

It starts to get more interesting, and potentially very useful, when we think about $x^{1/2}$. This has to satisfy $x^{1/2} \; x^{1/2} = x^{1/2+1/2} = x^1 = x$. So $x^{1/2}$, multiplied by itself, is x. The only number with this property is the square root of x. So $x^{1/2} = \sqrt{x}$. Similarly, $x^{1/3} = \sqrt[3]{x}$, the cube root. Continuing in this manner we can define $x^{p/q}$ for any fraction p/q. Then, using fractions to approximate real numbers, we can define x^a for any real a. And the equation $x^a x^b = x^{a+b}$ still holds.

It also follows that log $\sqrt{x} = \frac{1}{2}$ log x and log $\sqrt[3]{x} = \frac{1}{3}$ log x, so we can calculate square roots and cube roots easily using a table of logarithms. For example, to find the square root of a number we form its logarithm, divide by 2, and then work out which number has the result as its logarithm. For cube roots, do the same but divide by 3. Traditional methods for these problems were tedious and complicated. You can see why Napier showcased square and cube roots in the preface to his book.

As soon as complete tables of logarithms were available, they became an indispensable tool for scientists, engineers, surveyors, and navigators. They saved time, they saved effort, and they increased the likelihood that the answer was correct. Early on, astronomy was a major beneficiary, because astronomers routinely needed to perform long and difficult calculations. The French mathematician and astronomer Pierre Simon de Laplace said that the invention of logarithms 'reduces to a few days the labour of many months, doubles the life of the astronomer, and spares him the errors and disgust'. As the use of machinery in manufacturing grew, engineers started to make more and more use of mathematics – to design complex gears, analyse the stability of bridges and buildings, and construct cars, lorries, ships, and aeroplanes. Logarithms were a firm part of the school mathematics curriculum a few decades ago. And engineers carried what was in effect an analogue calculator for logarithms in their pockets, a physical representation of the basic equation for logarithms for on-the-spot use. They called it a slide rule, and they used it routinely in applications ranging from architecture to aircraft design.

The first slide rule was constructed by an English mathematician, William Oughtred, in 1630, using circular scales. He modified the design in 1632, by making the two rulers straight. This was the first slide rule. The idea is simple: when you place two rods end to end, their lengths add. If the rods are marked using a logarithmic scale, in which numbers are spaced according to their logarithms, then the corresponding numbers *multiply*. For instance, set the 1 on one rod against the 2 on the other. Then against any number x on the first rod, we find $2x$ on the second. So opposite 3 we find 6, and so on, see Figure 11. If the numbers are more complicated, say 2.67 and 3.51, we place 1 opposite 2.67 and read off whatever is opposite 3.59, namely 9.37. It's just as easy.

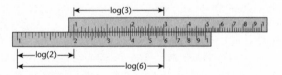

Fig 11 Multiplying 2 by 3 on a slide rule.

Engineers quickly developed fancy slide rules with trigonometric functions, square roots, log-log scales (logarithms of logarithms) to calculate powers, and so on. Eventually logarithms took a back seat to digital computers, but even now the logarithm still plays a huge role in science and technology, alongside its inseparable companion, the exponential function. For base-10 logarithms, this is the function 10^x; for natural logarithms, the function e^x, where $e = 2.71828$, approximately. In each pair, the two functions are inverse to each other. If you take a number, form its logarithm, and then form the exponential of that, you get back the number you started with.

Why do we need logarithms now that we have computers?

In 2011 a magnitude 9.0 earthquake just off the east coast of Japan caused a gigantic tsunami, which devastated a large populated area and killed around 25,000 people. On the coast was a nuclear power plant, Fukushima Dai-ichi (Fukushima number 1 power plant, to distinguish it from a second nuclear power plant situated nearby). It comprised six separate nuclear reactors: three were in operation when the tsunami struck; the other three had temporarily ceased operating and their fuel had been transferred to pools of water outside the reactors but inside the reactor buildings.

The tsunami overwhelmed the plant's defences, cutting the supply of electrical power. The three operating reactors (numbers 1, 2, and 3) were shut down as a safety measure, but their cooling systems were still needed to stop the fuel from melting. However, the tsunami also wrecked the emergency generators, which were intended to power the cooling system and other safety-critical systems. The next level of backup, batteries, quickly ran out of power. The cooling system stopped and the nuclear fuel in several reactors began to overheat. Improvising, the operators used fire engines to pump seawater into the three operating reactors, but this reacted with the zirconium cladding on the fuel rods to produce hydrogen. The build-up of hydrogen caused an explosion in the building housing Reactor 1. Reactors 2 and 3 soon suffered the same fate. The water in the pool of Reactor 4 drained out, leaving its fuel exposed. By the time the operators regained some semblance of control, at least one reactor containment vessel had cracked, and radiation was leaking out into the local environment. The Japanese authorities evacuated 200,000 people from the surrounding area because the radiation was well above normal safety limits. Six months later, the company operating the reactors, TEPCO, stated that the situation remained critical and much more work would be needed before the reactors could be considered fully under control, but claimed the leakage had been stopped.

I don't want to analyse the merits or otherwise of nuclear power here, but I do want to show how the logarithm answers a vital question: if you know how much radioactive material has been released, and of what kind, how long will it remain in the environment, where it could be hazardous?

Radioactive elements decay; that is, they turn into other elements through nuclear processes, emitting nuclear particles as they do so. It is these particles that constitute the radiation. The level of radioactivity falls away over time just as the temperature of a hot body falls when it cools:

exponentially. So, in appropriate units, which I won't discuss here, the level of radioactivity $N(t)$ at time t follows the equation

$$N(t) = N_0 e^{-kt}$$

where N_0 is the initial level and k is a constant depending on the element concerned. More precisely, it depends on which form, or isotope, of the element we are considering.

A convenient measure of the time radioactivity persists is the half-life, a concept first introduced in 1907. This is the time it takes for an initial level N_0 to drop to half that size. To calculate the half-life, we solve the equation

$$\tfrac{1}{2}N_0 = N_0 e^{-kt}$$

by taking logarithms of both sides. The result is

$$t = \frac{\log 2}{k} = \frac{0.6931}{k}$$

and we can work this out because k is known from experiments.

The half-life is a convenient way to assess how long the radiation will persist. Suppose that the half-life is one week, for instance. Then the original rate at which the material emits radiation halves after 1 week, is down to one quarter after 2 weeks, one eighth after 3 weeks, and so on. It takes 10 weeks to drop to one thousandth of its original level (actually 1/1024), and 20 weeks to drop to one millionth.

In accidents with conventional nuclear reactors, the most important radioactive products are iodine-131 (a radioactive isotope of iodine) and caesium-137 (a radioactive isotope of caesium). The first can cause thyroid cancer, because the thyroid gland concentrates iodine. The half-life of iodine-131 is only 8 days, so it causes little damage if the right medication is available, and its dangers decrease fairly rapidly unless it continues to leak. The standard treatment is to give people iodine tablets, which reduce the amount of the radioactive form that is taken up by the body, but the most effective remedy is to stop drinking contaminated milk.

Caesium-137 is very different: it has a half-life of 30 years. It takes about 200 years for the level of radioactivity to drop to one hundredth of its initial value, so it remains a hazard for a very long time. The main practical issue in a reactor accident is contamination of soil and buildings. Decontamination is to some extent feasible, but expensive. For example, the soil can be removed, carted away, and stored somewhere safe. But this creates huge amounts of low-level radioactive waste.

Radioactive decay is just one area of many in which Napier's and Briggs's logarithms continue to serve science and humanity. If you thumb through later chapters you will find them popping up in thermodynamics and information theory, for example. Even though fast computers have now made logarithms redundant for their original purpose, rapid calculations, they remain central to science for conceptual rather than computational reasons.

Another application of logarithms comes from studies of human perception: how we sense the world around us. The early pioneers of the psychophysics of perception made extensive studies of vision, hearing, and touch, and they turned up some intriguing mathematical regularities.

In the 1840s a German doctor, Ernst Weber, carried out experiments to determine how sensitive human perception is. His subjects were given weights to hold in their hands, and asked when they could tell that one weight felt heavier than another. Weber could then work out what the smallest detectable difference in weight was. Perhaps surprisingly, this difference (for a given experimental subject) was not a fixed amount. It depended on how heavy the weights being compared were. People didn't sense an absolute minimum difference – 50 grams, say. They sensed a *relative* minimum difference – 1% of the weights under comparison, say. That is, the smallest difference that the human senses can detect is proportional to the stimulus, the actual physical quantity.

In the 1850s Gustav Fechner rediscovered the same law, and recast it mathematically. This led him to an equation, which he called Weber's law, but nowadays it is usually called Fechner's law (or the Weber–Fechner law if you're a purist). It states that the perceived sensation is proportional to the *logarithm* of the stimulus. Experiments suggested that this law applies not only to our sense of weight but to vision and hearing as well. If we look at a light, the brightness that we perceive varies as the logarithm of the actual energy output. If one source is ten times as bright as another, then the difference we perceive is constant, however bright the two sources really are. The same goes for the loudness of sounds: a bang with ten times as much energy sounds a fixed amount louder.

The Weber–Fechner law is not totally accurate, but it's a good approximation. Evolution pretty much had to come up with something like a logarithmic scale, because the external world presents our senses with stimuli over a huge range of sizes. A noise may be little more than a mouse scuttling in the hedgerow, or it may be a clap of thunder; we need to be able to hear both. But the range of sound levels is so vast that no biological sensory device can respond in proportion to the energy generated by the sound. If an ear that could hear the mouse did that, then a thunderclap would destroy it. If it tuned the sound levels down so that the thunderclap produced a comfortable signal, it wouldn't be able to hear the mouse. The solution is to compress the energy levels into a comfortable range, and the logarithm does exactly that. Being sensitive to proportions rather than absolutes makes excellent sense, and makes for excellent senses.

Our standard unit for noise, the decibel, encapsulates the Weber–Fechner law in a definition. It measures not absolute noise, but relative noise. A mouse in the grass produces about 10 decibels. Normal conversation between people a metre apart takes place at 40–60 decibels. An electric mixer directs about 60 decibels at the person using it. The noise in a car, caused by engine and tyres, is 60–80 decibels. A jet airliner a hundred metres away produces 110–140 decibels, rising to 150 at thirty metres. A vuvuzela (the annoying plastic trumpet-like instrument widely heard during the football World Cup in 2010 and brought home as souvenirs by misguided fans) generates 120 decibels at one metre; a military stun grenade produces up to 180 decibels.

Scales like these are widely encountered because they have a safety aspect. The level at which sound can potentially cause hearing damage is about 120 decibels. Please throw away your vuvuzela.

3 Ghosts of departed quantities
Calculus

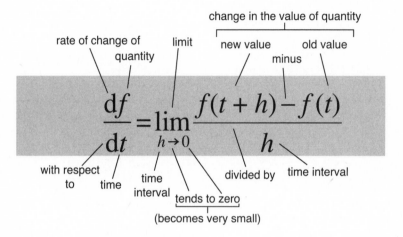

change in the value of quantity

rate of change of quantity   limit   new value   old value

minus

$$\frac{df}{dt} = \lim_{h \to 0} \frac{f(t+h) - f(t)}{h}$$

with respect to   time   time interval   divided by   time interval

tends to zero
(becomes very small)

What does it say?

To find the instantaneous rate of change of a quantity that varies with (say) time, calculate how its value changes over a short time interval and divide by the time concerned. Then let that interval become arbitrarily small.

Why is that important?

It provides a rigorous basis for calculus, the main way scientists model the natural world.

What did it lead to?

Calculation of tangents and areas. Formulas for volumes of solids and lengths of curves. Newton's laws of motion, differential equations. The laws of conservation of energy and momentum. Most of mathematical physics.

I n 1665 Charles II was king of England and his capital city, London, was a sprawling metropolis of half a million people. The arts flourished, and science was in the early stages of an ever-accelerating ascendancy. The Royal Society, perhaps the oldest scientific society now in existence, had been founded five years earlier, and Charles had granted it a royal charter. The rich lived in impressive houses, and commerce was thriving, but the poor were crammed into narrow streets overshadowed by ramshackle buildings that jutted out ever further as they rose, storey by storey. Sanitation was inadequate; rats and other vermin were everywhere. By the end of 1666, one fifth of London's population had been killed by bubonic plague, spread first by rats and then by people. It was the worst disaster in the capital's history, and the same tragedy played out all over Europe and North Africa. The king departed in haste for the more sanitary countryside of Oxfordshire, returning early in 1666. No one knew what caused plague, and the city authorities tried everything – burning fires continually to cleanse the air, burning anything that gave off a strong smell, burying the dead quickly in pits. They killed many dogs and cats, which ironically removed two controls on the rat population.

During those two years, an obscure and unassuming undergraduate at Trinity College, Cambridge, completed his studies. Hoping to avoid the plague, he returned to the house of his birth, from which his mother managed a farm. His father had died shortly before he was born, and he had been brought up by his maternal grandmother. Perhaps inspired by rural peace and quiet, or lacking anything better to do with his time, the young man thought about science and mathematics. Later he wrote: 'In those days I was in the prime of my life for invention, and minded mathematics and [natural] philosophy more than at any other time since.' His researches led him to understand the importance of the inverse square law of gravity, an idea that had been hanging around ineffectually for at least 50 years. He worked out a practical method for solving problems in calculus, another concept that was in the air but had not been formulated in any generality. And he discovered that white sunlight is composed of many different colours – all the colours of the rainbow.

When the plague died down, he told no one about the discoveries he had made. He returned to Cambridge, took a master's degree, and became a fellow at Trinity. Elected to the Lucasian Chair of Mathematics, he finally began to publish his ideas and to develop new ones.

The young man was Isaac Newton. His discoveries created a revolution in science, bringing about a world that Charles II would never have believed could exist: buildings with more than a hundred floors, horseless carriages doing 80 mph along the M6 motorway while the driver listens to music using a magic disc made from a strange glasslike material, heavier-than-air flying machines that cross the Atlantic in six hours, colour pictures that move, and boxes you carry in your pocket that talk to the other side of the world...

Previously, Galileo Galilei, Johannes Kepler, and others had turned up the corner of nature's rug and seen a few of the wonders concealed beneath it. Now Newton cast the rug aside. Not only did he reveal that the universe has secret patterns, laws of nature; he also provided mathematical tools to express those laws precisely and to deduce their consequences. The system of the world was mathematical; the heart of God's creation was a soulless clockwork universe.

The world view of humanity did not suddenly switch from religious to secular. It still has not done so completely, and probably never will. But after Newton published his *Philosophiæ Naturalis Principia Mathematica* ('Mathematical Principles of Natural Philosophy') the 'System of the World' – the book's subtitle – was no longer solely the province of organised religion. Even so, Newton was not the first modern scientist; he had a mystical side too, devoting years of his life to alchemy and religious speculation. In notes for a lecture[1] the economist John Maynard Keynes, also a Newtonian scholar, wrote:

Newton was not the first of the age of reason. He was the last of the magicians, the last of the Babylonians and Sumerians, the last great mind which looked out on the visible and intellectual world with the same eyes as those who began to build our intellectual inheritance rather less than 10,000 years ago. Isaac Newton, a posthumous child born with no father on Christmas Day, 1642, was the last wonderchild to whom the Magi could do sincere and appropriate homage.

Today we mostly ignore Newton's mystic aspect, and remember him for his scientific and mathematical achievements. Paramount among them are his realisation that nature obeys mathematical laws and his invention of calculus, the main way we now express those laws and derive their consequences. The German mathematician and philosopher Gottfried Wilhelm Leibniz also developed calculus, more or less independently, at much the same time, but he did little with it. Newton used calculus to understand the universe, though he kept it under wraps in his published work, recasting it in classical geometric language. He was a transitional figure who moved humanity away from a mystical, medieval outlook and ushered in the modern rational world view. After Newton, scientists consciously recognised that the universe has deep mathematical patterns, *and* were equipped with powerful techniques to exploit that insight.

The calculus did not arise 'out of the blue'. It came from questions in both pure and applied mathematics, and its antecedents can be traced back to Archimedes. Newton himself famously remarked, 'If I have seen a little further it is by standing on the shoulders of giants.'[2] Paramount among those giants were John Wallis, Pierre de Fermat, Galileo, and Kepler. Wallis developed a precursor to calculus in his 1656 *Arithmetica Infinitorum* ('Arithmetic of the Infinite'). Fermat's 1679 *De Tangentibus Linearum Curvarum* ('On Tangents to Curved Lines') presented a method for finding tangents to curves, a problem intimately related to calculus. Kepler formulated three basic laws of planetary motion, which led Newton to his law of gravity, the subject of the next chapter. Galileo made big advances in astronomy, but he also investigated mathematical aspects of nature down on the ground, publishing his discoveries in *De Motu* ('On Motion') in 1590. He investigated how a falling body moves, finding an elegant mathematical pattern. Newton developed this hint into three general laws of motion.

To understand Galileo's pattern we need two everyday concepts from mechanics: velocity and acceleration. Velocity is how fast something is moving, and in which direction. If we ignore the direction, we get the body's speed. Acceleration is a change in velocity, which usually involves a change in speed (an exception arises when the speed remains the same but the direction changes). In everyday life we use acceleration to mean speeding up and deceleration for slowing down, but in mechanics both changes are accelerations: the first positive, the second negative. When we drive along a road the speed of the car is displayed on the speedometer – it might, for instance, be 50 mph. The direction is whichever way the car is pointing. When we put our foot down, the car accelerates and the speed increases; when we stamp on the brakes, the car decelerates – negative acceleration.

If the car is moving at a fixed speed, it's easy to work out what that speed is. The abbreviation mph gives it away: miles per hour. If the car travels 50 miles in 1 hour, we divide the distance by the time, and that's the speed. We don't need to drive for an hour: if the car goes 5 miles in 6 minutes, both distance and time are divided by 10, and their ratio is still 50 mph. In short,

speed = distance travelled divided by time taken.

In the same way, a fixed rate of acceleration is given by

acceleration = change in speed divided by time taken.

This all seems straightforward, but conceptual difficulties arise when the speed or acceleration is not fixed. And they can't both be constant, because constant (and nonzero) acceleration implies a changing speed. Suppose you drive along a country lane, speeding up on the straights, slowing for the corners. Your speed keeps changing, and so does your acceleration. How can we work them out at any given instant of time? The pragmatic answer is to take a short interval of time, say a second. Then your instantaneous speed at (say) 11.30 am is the distance you travel between that moment and one second later, divided by one second. The same goes for instantaneous acceleration.

Except ... that's not quite your *instantaneous* speed. It's really an average speed, over a one-second interval of time. There are circumstances in which one second is a *huge* length of time – a guitar string playing middle C vibrates 440 times every second; average its motion over an entire second and you'll think it's standing still. The answer is to consider a shorter interval of time – one ten thousandth of a second, perhaps. But this still doesn't capture instantaneous speed. Visible light vibrates one quadrillion (10^{15}) times every second, so the appropriate time interval is less than one quadrillionth of a second. And even then ... well, to be pedantic, that's still not an *instant*. Pursuing this line of thought, it seems to be necessary to use an interval of time that is shorter than any other interval. But the only number like that is 0, and that's useless, because now the distance travelled is also 0, and 0/0 is meaningless.

Early pioneers ignored these issues and took a pragmatic view. Once the probable error in your measurements exceeds the increased precision you would theoretically get by using smaller intervals of time, there's no point in doing so. The clocks in Galileo's day were very inaccurate, so he measured time by humming tunes to himself – a trained musician can subdivide a note into very short intervals. Even then, timing a falling body is tricky, so Galileo hit on the trick of slowing the motion down by rolling balls down an inclined slope. Then he observed the position of the ball at successive intervals of time. What he found (I'm simplifying the numbers to make the pattern clear, but it's the same pattern) is that for times 0, 1, 2, 3, 4, 5, 6, ... these positions were

$$0 \quad 1 \quad 4 \quad 9 \quad 16 \quad 25 \quad 36$$

The distance was (proportional to) the square of the time. What about the speeds? Averaged over successive intervals, these were the differences

$$1 \quad 3 \quad 5 \quad 7 \quad 9 \quad 11$$

between the successive squares. In each interval, other than the first, the average speed increased by 2 units. It's a striking pattern – all the more so to Galileo when he dug something very similar out of dozens of measurements with balls of many different masses on slopes with many different inclinations.

From these experiments and the observed pattern, Galileo deduced something wonderful. The path of a falling body, or one thrown into the air, such as a cannonball, is a parabola. This is a U-shaped curve, known to the ancient Greeks. (The U is upside down in this case. I'm ignoring air resistance, which changes the shape: it didn't have much effect on Galileo's rolling balls.) Kepler encountered a related curve, the ellipse, in his analysis of planetary orbits: this must have seemed significant to Newton too, but that story must wait until the next chapter.

With only this particular series of experiments to go on, it's not clear what general principles underlie Galileo's pattern. Newton realised that the source of the pattern is rates of change. Velocity is the rate at which position changes with respect to time; acceleration is the rate at which velocity changes with respect to time. In Galileo's observations, position varied according to the square of time, velocity varied linearly, and acceleration didn't vary at all. Newton realised that in order to gain a deeper understanding of Galileo's patterns, and what they meant for our view of nature, he had to come to grips with instantaneous rates of change. When he did, out popped calculus.

You might expect an idea as important as calculus to be announced with a fanfare of trumpets and parades through the streets. However, it takes time for the significance of novel ideas to sink in and to be appreciated, and so it was with calculus. Newton's work on the topic dates from 1671 or earlier, when he wrote *The Method of Fluxions and Infinite Series*. We are unsure of the date because the book was not published until 1736, nearly a decade after his death. Several other manuscripts by Newton also refer to ideas that we now recognise as differential and integral calculus, the two main branches of the subject. Leibniz's notebooks show that he obtained his first significant results in calculus in 1675, but he published nothing on the topic until 1684.

After Newton had risen to scientific prominence, long after both men had worked out the basics of calculus, some of Newton's friends sparked a largely pointless but heated controversy about priority, accusing Leibniz of plagiarising Newton's unpublished manuscripts. A few mathematicians from continental Europe responded with counter-claims of plagiarism by Newton. English and continental mathematicians were scarcely on speaking terms for a century, which caused huge damage to English mathematicians, but none whatsoever to the continental ones. They developed calculus into a central tool of mathematical physics while their English counterparts were seething about insults to Newton instead of exploiting insights from Newton. The story is tangled and still subject to scholarly disputation by historians of science, but broadly speaking it seems that Newton and Leibniz discovered the basic ideas of calculus independently – at least, as independently as their common mathematical and scientific culture permitted.

Leibniz's notation differs from Newton's, but the underlying ideas are more or less identical. The intuition behind them, however, is different. Leibniz's approach was a formal one, manipulating algebraic symbols. Newton had a physical model at the back of his mind, in which the function under consideration was a physical quantity that varies with time. This is where his curious term 'fluxion' comes from – something that flows as time passes.

Newton's method can be illustrated using an example: a quantity y that is the square x^2 of another quantity x. (This is the pattern that Galileo found for a rolling ball: its position is proportional to the square of the time that has elapsed. So there y would be position and x time. The usual symbol for time is t, but the standard coordinate system in the plane uses x and y.) Start by introducing a new quantity o, denoting a small change in x. The corresponding change in y is the difference

$$(x+o)^2 - x^2$$

which simplifies to $2xo+o^2$. The rate of change (averaged over a small interval of length o, as x increases to $x+o$) is therefore

$$\frac{2xo+o^2}{o} = 2x+o$$

This depends on o, which is only to be expected since we are averaging the rate of change over a nonzero interval. However, if o becomes smaller and smaller, 'flowing towards' zero, the rate of change $2x+o$ gets closer and closer to $2x$. This does not depend on o, and it gives the instantaneous rate of change at x.

Leibniz performed essentially the same calculation, replacing o by dx ('small difference in x'), and defining dy to be the corresponding small change in y. When a variable y depends on another variable x, the rate of change of y with respect to x is called the derivative of y. Newton wrote the derivative of y by placing a dot above it: $\dot{y}$. Leibniz wrote $\frac{dy}{dx}$. For higher derivatives, Newton used more dots, while Leibniz wrote things like $\frac{d^2y}{dx^2}$. Today we say that y is a *function* of x and write $y=f(x)$, but this concept existed only in rudimentary form at the time. We either use Leibniz's notation, or a variant of Newton's in which the dot is replaced by a dash, which is easier to print: y', y''. We also write $f'(x)$ and $f''(x)$ to emphasise that the derivatives are themselves functions. Calculating the derivative is called differentiation.

Integral calculus – finding areas – turns out to be the inverse of differential calculus – finding slopes. To see why, imagine adding a thin slice on the end of the shaded area of Figure 12. This slice is very close to a long thin rectangle, of width o and height y. Its area is therefore very close to oy. The rate at which the area changes, with respect to x, is the ratio oy/o, which equals y. So the derivative of the area is the original function. Both Newton and Leibniz understood that the way to calculate the area, a process called integration, is the reverse of differentiation in this sense. Leibniz first wrote the integral using the symbol omn., short for *omnia*, or 'sum', in Latin. Later he changed this to ∫, an old-fashioned long s, also standing for 'sum'. Newton had no systematic notation for the integral.

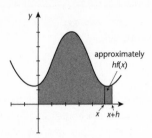

Fig 12 Adding a thin slice to the area beneath the curve $y = f(x)$.

Newton did make one crucial advance, however. Wallis had calculated the derivative of any power x^a: it is ax^{a-1}. So the derivatives of x^3, x^4, x^5 are $3x^2$, $4x^3$, $5x^4$, for example. He had extended this result to any polynomial – a finite combination of powers, such as $3x^7 - 25x^4 + x^2 - 3$. The trick is to consider each power separately, find the corresponding derivatives, and combine them in the same manner. Newton noticed that the same method worked for infinite series, expressions involving infinitely many powers of the variable. This let him perform the operations of calculus on many other expressions, more complicated than polynomials.

Given the close correspondence between the two versions of calculus, differing mainly in unimportant features of the notation, it is easy to see how a priority dispute might have arisen. However, the basic idea is a fairly direct formulation of the underlying question, so it is also easy to see how Newton and Leibniz could have arrived at their versions independently, despite the similarities. In any case, Fermat and Wallis had beaten them both to many of their results. The dispute was pointless.

A more fruitful controversy concerned the logical structure of calculus, or more precisely, the illogical structure of calculus. A leading critic was the Anglo-Irish philosopher George Berkeley, Bishop of Cloyne. Berkeley had a religious agenda; he felt that the materialist view of the world that was developing from Newton's work represented God as a detached creator who stood back from his creation as soon as it got going and thereafter left it to its own devices, quite unlike the personal, immanent God of Christian belief. So he attacked logical inconsistencies in the foundations of calculus, presumably hoping to discredit the resulting science. His attack had no discernible effect on the progress of mathematical physics, for a straightforward reason: the results obtained using calculus shed so much insight into nature, and agreed so well with experiment, that the logical foundations seemed unimportant. Even today, physicists still take this view: if it works, who cares about logical hair-splitting?

Berkeley argued that it makes no logical sense to maintain that a small quantity (Newton's o, Leibniz's dx) is nonzero for most of a calculation, and then to set it to zero, if you have previously divided both the numerator and the denominator of a fraction by that very quantity. Division by zero is not an acceptable operation in arithmetic, because it has no unambiguous meaning. For example, $0 \times 1 = 0 \times 2$, since both are 0, but if we divide both sides of this equation by 0 we get $1 = 2$, which is false.[3] Berkeley published his criticisms in 1734 in a pamphlet *The Analyst, a Discourse Addressed to an Infidel Mathematician*.

Newton had, in fact, attempted to sort out the logic, by appealing to a physical analogy. He saw o not as a fixed quantity, but as something that *flowed* – varied with time – getting closer and closer to zero without ever actually getting there. The derivative was also defined by a quantity that flowed: the ratio of the change in y to that of x. This ratio also flowed towards something, but never got there; that something was the instantaneous rate of change – the derivative of y with respect to x. Berkeley dismissed this idea as the 'ghost of a departed quantity'.

Leibniz too had a persistent critic, the geometer Bernard Nieuwentijt, who put his criticisms into print in 1694 and 1695. Leibniz had not helped his case by trying to justify his method in terms of 'infinitesimals', a term open to misinterpretation. However, he did explain that what he meant by this term was not a fixed nonzero quantity that can be arbitrarily small (which makes no logical sense) but a variable nonzero quantity that can *become* arbitrarily small. Newton's and Leibniz's defences were essentially identical. To their opponents, both must have sounded like verbal trickery.

Fortunately, the physicists and mathematicians of the day did not wait for the logical foundations of calculus to be sorted out before they applied it to the frontiers of science. They had an alternative way to make sure they were doing something sensible: comparison with observations and experiments. Newton himself invented calculus for precisely this purpose. He derived laws for how bodies move when a force is applied to them, and combined these with a law for the force exerted by gravity to explain many riddles about the planets and other bodies of the Solar System. His law of gravity is such a pivotal equation in physics and astronomy that it deserves, and gets, a chapter of its own (the next one). His law of motion – strictly, a system of three laws, one of which contained most of the mathematical content – led fairly directly to calculus.

Ironically, when Newton published these laws and their scientific applications in his *Principia*, he eliminated all traces of calculus and replaced it by classical geometric arguments. He probably thought that geometry would be more acceptable to his intended audience, and if he did, he was almost certainly right. However, many of his geometric proofs are either motivated by calculus, or depend on the use of calculus techniques to determine the correct answers, upon which the strategy of the geometric proof relies. This is especially clear, to modern eyes, in his treatment of what he called 'generated quantities' in Book II of *Principia*. These are quantities that increase or decrease by 'continual motion or flux', the fluxions of his unpublished book. Today we would call them continuous (indeed differentiable) functions. In place of explicit operations of the calculus, Newton substituted a geometric method of 'prime and ultimate ratios'. His opening lemma (the name given to an auxiliary mathematical result that is used repeatedly but has no intrinsic interest in its own right) gives the game away, because it *defines* equality of these flowing quantities like this:

> Quantities, and the ratios of quantities, which in any finite time converge continually to equality, and before the end of that time approach nearer to each other than by any given difference, become ultimately equal.

In *Never at Rest*, Newton's biographer Richard Westfall explains how radical and novel this lemma was: 'Whatever the language, the concept ... was thoroughly modern; classical geometry had contained nothing like it.'[4] Newton's contemporaries must have struggled to figure out what Newton was getting at. Berkeley presumably never did, because – as we will shortly see – it contains the basic idea needed to dispose of his objection.

Calculus, then, was playing an influential role behind the scenes of the *Principia*, but it made no appearance on stage. As soon as calculus peeped out from behind the curtains, however, Newton's intellectual successors quickly reverse-engineered his thought processes. They rephrased his main ideas in the language of calculus, because this provided a more natural and more powerful framework, and set out to conquer the scientific world.

The clue was already visible in Newton's laws of motion. The question that led Newton to these laws was a philosophical one: what causes a body to move, or to change its state of motion? The classical answer was Aristotle's: a body moves because a force is applied to it, and this affects its velocity. Aristotle also stated that in order to keep a body moving, the force must continue to be applied. You can test Aristotle's statements by placing a book or similar object on a table. If you push the book, it starts to move, and if you keep pushing with much the same force it continues to slide over the table at a roughly constant velocity. If you stop pushing, the book stops moving. So Aristotle's views seem to agree with experiment. However, the agreement is superficial, because the push is not the only force that acts on the book. There is also friction with the surface of the table. Moreover, the faster the book moves, the greater the friction becomes – at least, while the book's velocity remains reasonably small. When the book is moving steadily across the table, propelled by a steady force, the frictional resistance cancels out the applied force, and the total force acting on the body is actually zero.

Newton, following earlier ideas of Galileo and Descartes, realised this. The resulting theory of motion is very different from Aristotle's. Newton's three laws are:

First law. Every body continues in its state of rest, or of uniform motion in a right [straight] line, unless it is compelled to change that state by forces impressed upon it.

Second law. The change of motion is proportional to the motive power impressed, and is made in the direction of the right line in which that force is impressed. (The constant of proportionality is the reciprocal of the body's mass; that is, 1 divided by that mass.)

Third law. To every action there is always opposed an equal reaction.

The first law explicitly contradicts Aristotle. The third law says that if you push something, it pushes back. The second law is where calculus comes in. By 'change of motion' Newton meant the rate at which the body's velocity changes: its acceleration. This is the derivative of velocity with respect to time, and the second derivative of position. So Newton's second law of motion specifies the relation between a body's position, and the forces that act on it, in the form of a *differential equation*:

$$\text{second derivative of position} = \text{force/mass}$$

To find the position itself, we have to solve this equation, deducing the position from its second derivative.

This line of thought leads to a simple explanation of Galileo's observations of a rolling ball. The crucial point is that the acceleration of the ball is *constant*. I stated this previously, using a rough-and-ready calculation applied at discrete intervals of time; now we can do it properly, allowing time to vary continuously. The constant is related to the force of gravity and the angle of the slope, but here we don't need that much detail. Suppose that the constant acceleration is a. Integrating the corresponding function, the velocity down the slope at time t is $at+b$, where b is the velocity at time zero. Integrating again, the position down the slope is $\frac{1}{2}at^2+bt+c$, where c is the position at time zero. In the special case $a=2$, $b=0$, $c=0$ the successive positions fit my simplified example: the position at time t is t^2. A similar analysis recovers Galileo's major result: the path of a projectile is a parabola.

Newton's laws of motion did not just provide a way to calculate how bodies move. They led to deep and general physical principles. Paramount among these are 'conservation laws', telling us that when a system of bodies, no matter how complicated, moves, certain features of that system *do not change*. Amid the tumult of the motion, a few things remain serenely unaffected. Three of these conserved quantities are energy, momentum, and angular momentum.

Energy can be defined as the capacity to do work. When a body is raised to a certain height, against the (constant) force of gravity, the work done to put it there is proportional to the body's mass, the force of gravity, and the height to which it is raised. Conversely, if we then let the body go, it can perform the same amount of work when it falls back to its original height. This type of energy is called *potential energy*.

On its own, potential energy would not be terribly interesting, but there is a beautiful mathematical consequence of Newton's second law of motion leading to a second kind of energy: *kinetic energy*. As a body moves, both its potential energy and its kinetic energy change. But the change in one exactly compensates for the change in the other. As the body descends under gravity, it speeds up. Newton's law allows us to calculate how its velocity changes with height. It turns out that the decrease in potential energy is exactly equal to half the mass times the square of the velocity. If we give that quantity a name – kinetic energy – then the total energy, potential plus kinetic, is conserved. This mathematical consequence of Newton's laws proves that perpetual motion machines are impossible: no mechanical device can keep going indefinitely and do work without some external input of energy.

Physically, potential and kinetic energy seem to be two different things; mathematically, we can trade one for the other. It is as if motion somehow converts potential energy into kinetic. 'Energy', as a term applicable to both, is a convenient abstraction, carefully defined so that it is conserved. As an analogy, travellers can convert pounds into dollars. Currency exchanges have tables of exchange rates, asserting that, say, 1 pound is of equal value to 1.4693 dollars. They also deduct a sum of money for themselves. Subject to technicalities of bank charges and so on, the total monetary value involved in the transaction is supposed to balance out: the traveller gets exactly the amount in dollars that corresponds to their original sum in pounds, minus various deductions. However, there isn't a physical *thing* built into banknotes that somehow gets swapped out of a pound note into a dollar note and some coins. What gets swapped is the human convention that these particular items have monetary value.

Energy is a new kind of 'physical' quantity. From a Newtonian viewpoint, quantities such as position, time, velocity, acceleration, and mass have direct physical interpretations. You can measure position with a ruler, time with a clock, velocity and acceleration using both pieces of apparatus, and mass with a balance. But you don't measure energy using an energy meter. Agreed, you can measure certain specific *types* of energy. Potential energy is proportional to height, so a ruler will suffice if you know the force of gravity. Kinetic energy is half the mass times the square of the velocity: use a balance and a speedometer. But *energy*, as a concept, is not so much a physical thing as a convenient fiction that helps to balance the mechanical books.

Momentum, the second conserved quantity, is a simple concept: mass times velocity. It comes into play when there are several bodies. An important example is a rocket; here one body is the rocket and the other is its fuel. As fuel is expelled by the engine, conservation of momentum implies that the rocket must move in the opposite direction. This is how a rocket works in a vacuum.

Angular momentum is similar, but it relates to spin rather than velocity. It is also central to rocketry, indeed the whole of mechanics, terrestrial or celestial. One of the biggest puzzles about the Moon is its large angular momentum. The current theory is that the Moon was splashed off when a Mars-sized planet hit the Earth about 4.5 billion years ago. This explains the angular momentum, and until recently was generally accepted, but it now seems that the Moon has too much water in its rocks. Such an impact should have boiled a lot of the water away.[5] Whatever the eventual outcome, angular momentum is of central importance here.

Calculus works. It solves problems in physics and geometry, getting the right answers. It even leads to new and fundamental physical concepts like energy and momentum. But that doesn't answer Bishop Berkeley's objection. Calculus has to work as mathematics, not just agree with physics. Both Newton and Leibniz understood that o or dx cannot be both zero and nonzero. Newton tired to escape from the logical trap by employing the physical image of a fluxion. Leibniz talked of infinitesimals. Both referred to quantities that approach zero without ever getting there – but what are these things? Ironically, Berkeley's gibe about 'ghosts of departed quantities' comes close to resolving the issue, but what he failed to take account of – and what both Newton and Leibniz emphasised – was *how* the quantities departed. Make them depart in the right way and you can leave a perfectly well-formed ghost. If either Newton or Leibniz had framed their intuition in rigorous mathematical language, Berkeley might have understood what they were getting at.

The central question is one that Newton failed to answer explicitly because it seemed obvious. Recall that in the example where $y = x^2$, Newton obtained the derivative as $2x + o$, and then asserted that as o flows towards zero, $2x + o$ flows towards $2x$. This may seem obvious, but we can't set $o = 0$ to prove it. It is true that *we get the right result by doing that*, but this is a red herring.[6] In *Principia* Newton slid round this issue altogether, replacing $2x + o$ by his 'prime ratio' and $2x$ by his 'ultimate ratio'. But the real key to progress is to tackle the issue head on. How do we *know* that the closer o approaches zero, the closer $2x + o$ approaches $2x$? It may seem a rather pedantic point, but if I'd used more complicated examples the correct answer might not seem so plausible.

When mathematicians returned to the logic of calculus, they realised that this apparently simple question was the heart of the matter. When we say that o approaches zero, we mean that given any nonzero positive number, o can be chosen to be smaller than that number. (This is obvious: let o be half that number, for instance.) Similarly, when we say that $2x + o$ approaches $2x$, we mean that the difference approaches zero, in the previous sense. Since the difference happens to be o itself in this case, that's even more obvious: whatever 'approaches zero' means, clearly o approaches zero when o approaches zero. A more complicated function than the square would require a more complicated analysis.

The answer to this key question is to state the process in formal mathematical terms, avoiding ideas of 'flow' altogether. This breakthrough came about through the work of the Bohemian mathematician and theologian Bernard Bolzano and the German mathematician Karl

Weierstrass. Bolzano's work dates from 1816, but it was not appreciated until about 1870 when Weierstrass extended the formulation to complex functions. Their answer to Berkeley was the concept of a limit. I'll state the definition in words and leave the symbolic version to the Notes.[7] Say that a function $f(h)$ of a variable h tends to a limit L as h tends to zero if, given any positive nonzero number, the difference between $f(h)$ and L can be made smaller than that number by choosing sufficiently small nonzero values of h. In symbols,

$$\lim_{h \to 0} f(h) = L$$

The idea at the heart of calculus is to approximate the rate of change of a function over a small interval h, and then take the limit as h tends to zero. For a general function $y = f(x)$ this procedure leads to the equation that decorates the opening of this chapter, but using a general variable x instead of time:

$$f'(x) = \lim_{h \to 0} \frac{f(x+h) - f(x)}{h}$$

In the numerator we see the change in f; the denominator is the change in x. This equation defines the derivative $f'(x)$ uniquely, provided the limit exists. That has to be proved for any function under consideration: the limit does exist for most of the standard functions – squares, cubes, higher powers, logarithms, exponentials, trigonometric functions.

Nowhere in the calculation do we ever divide by zero, because we never set $h = 0$. Moreover, nothing here actually flows. What matters is the range of values that h can assume, not how it moves through that range. So Berkeley's sarcastic characterisation is actually spot on. The limit L is the ghost of the departed quantity – my h, Newton's o. But the manner of the quantity's departure – *approaching* zero, not *reaching* it – leads to a perfectly sensible and logically well-defined ghost.

Calculus now had a sound logical basis. It deserved, and acquired, a new name to reflect its new status: analysis.

It is no more possible to list all the ways that calculus can be applied than it is to list everything in the world that depends on using a screwdriver. On a simple computational level, applications of calculus include finding lengths of curves, areas of surfaces and complicated shapes, volumes of solids, maximum and minimum values, and centres of mass. In conjunction with the laws of mechanics, calculus tells us how to work out the trajectory of a space rocket, the stresses in rock at a subduction zone that might produce an earthquake, the way a building will vibrate if an earthquake hits, the way a car bounces up and down on its suspension, the time it takes a bacterial infection to spread, the way a surgical wound heals, and the forces that act on a suspension bridge in a high wind.

Many of these applications stem from the deep structure of Newton's laws: they are models of nature stated as differential equations. These are equations involving derivatives of an unknown function, and techniques from calculus are needed to solve them. I will say no more here, because every chapter from Chapter 8 onwards involves calculus explicitly, mainly in the guise of differential equations. The sole exception is Chapter 15 on information theory, and even there other developments that I don't mention also involve calculus. Like the screwdriver, calculus is simply an indispensable tool in the engineer's and scientist's toolkits. More than any other mathematical technique, it has created the modern world.

4 The system of the world
Newton's Law of Gravity

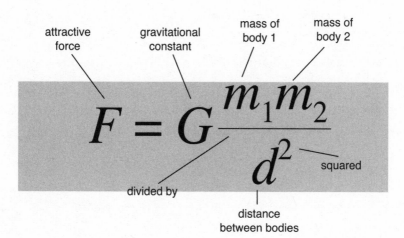

attractive force · gravitational constant · mass of body 1 · mass of body 2

$$F = G\frac{m_1 m_2}{d^2}$$

divided by · squared · distance between bodies

What does it say?

It determines the force of gravitational attraction between two bodies in terms of their masses and the distance between them.

Why is that important?

It can be applied to any system of bodies interacting through the force of gravity, such as the Solar System. It tells us that their motion is determined by a simple mathematical law.

What did it lead to?

Accurate prediction of eclipses, planetary orbits, the return of comets, the rotation of galaxies. Artificial satellites, surveys of the Earth, the Hubble telescope, observations of solar flares. Interplanetary probes, Mars rovers, satellite communications and television, the Global Positioning System.

Newton's laws of motion capture the relationship between the forces that act on a body and how it moves in response to those forces. Calculus provides mathematical techniques for solving the resulting equations. One further ingredient is needed to apply the laws: specifying the forces. The most ambitious aspect of Newton's *Principia* was to do precisely that for the bodies of the Solar System – the Sun, planets, moons, asteroids, and comets. Newton's law of gravitation synthesised, in one simple mathematical formula, millennia of astronomical observations and theories. It explained many puzzling features of planetary motion, and made it possible to predict the future movements of the Solar System with great accuracy. Einstein's theory of general relativity eventually superseded the Newtonian theory of gravity, as far as fundamental physics is concerned, but for almost all practical purposes the simpler Newtonian approach still reigns supreme. Today the world's space agencies, such as NASA and ESA, still use Newton's laws of motion and gravitation to work out the most effective trajectories for spacecraft.

It was Newton's law of gravitation, above all else, that justified his subtitle: *The System of the World*. This law demonstrated the enormous power of mathematics to find hidden patterns in nature and to reveal hidden simplicities behind the world's complexities. And in time, as mathematicians and astronomers asked harder questions, to reveal the hidden complexities implicit in Newton's simple law. To appreciate what Newton achieved, we must first go back in time, to see how previous cultures viewed the stars and planets.

Humans have been watching the night sky since the dawn of history. Their initial impression would have been a random scattering of bright points of light, but they would soon have noticed that across this background the glowing orb of the Moon traced a regular path, changing shape as it did so. They would also have seen that most of those tiny bright specks of light remain in the same relative patterns, which we now call constellations. Stars move across the night sky, but they move as a single rigid unit, as if the constellations are painted on the inside of a gigantic, rotating bowl.[1] However, a small number of stars behave quite differently: they seem to wander around the sky. Their paths are quite complicated, and some appear to loop back on themselves from time to time. These are the planets, a word that comes from the Greek for 'wanderer'. The ancients recognised five of them, now called Mercury, Venus, Mars, Jupiter, and Saturn. They move relative to the fixed stars at different speeds, with Saturn being the slowest.

Other celestial phenomena were even more puzzling. From time to time a comet would appear, as if from nowhere, trailing a long, curved tail. 'Shooting stars' would seem to fall from the heavens, as if they had become detached from their supporting bowl. It is no wonder that early humans attributed the irregularities of the heavens to the caprices of supernatural beings.

The regularities could be summed up in terms so obvious that few would ever dream of disputing them. The Sun, stars, and planets revolve around a stationary Earth. That's what it looks like, that's what it feels like, so that's how it must be. To the ancients, the cosmos was geocentric – Earth-centred. One lone voice disputed the obvious: Aristarchus of Samos. Using geometrical principles and observations, Aristarchus calculated the sizes of the Earth, the Sun, and the Moon. Around 270 BC he put forward the first heliocentric theory: the Earth and planets revolve round the Sun. His theory quickly fell out of favour and was not revived for nearly 2000 years.

By the time of Ptolemy, a Roman who lived in Egypt around 120 AD, the planets had been tamed. Their movements were not capricious, but predictable. Ptolemy's *Almagest* ('Great Treatise') proposed that we live in a geocentric universe in which everything literally revolves around humanity in complex combinations of circles called epicycles, supported by giant crystal spheres. His theory was wrong, but the motions that it predicted were sufficiently accurate for the errors to remain undetected for centuries. Ptolemy's system had an additional philosophical attraction: it represented the cosmos in terms of perfect geometric figures – spheres and circles. It continued the Pythagorean tradition. In Europe, the Ptolemaic theory remained unchallenged for 1400 years.

While Europe dawdled, new scientific advances were being made elsewhere, especially in Arabia, China, and India. In 499 the Indian astronomer Aryabhata put forward a mathematical model of the Solar System in which the Earth spun on its axis and the periods of planetary orbits were stated relative to the position of the Sun. In the Islamic world,

Alhazen wrote a stinging criticism of the Ptolemaic theory, though this was probably not focused on its geocentric nature. Around 1000 Abu Rayhan Biruni gave serious consideration to the possibility of a heliocentric Solar System, with the Earth spinning on its axis, but eventually plumped for the orthodoxy of the time, a stationary Earth. Around 1300, Najm al-Din al-Qazwini al-Katibi proposed a heliocentric theory, but soon changed his mind.

The big breakthrough came with the work of Nicolaus Copernicus, published in 1543 as *De Revolutionibus Orbium Coelestium* ('On the Revolutions of the Celestial Spheres'). There is evidence, notably the occurrence of almost identical diagrams labelled with the same letters, to suggest that Copernicus was, to say the least, influenced by al-Katibi, but he went much further. He set out an explicitly heliocentric system, argued that it fitted the observations better and more economically than Ptolemy's geocentric theory did, and laid out some of the philosophical implications. Paramount among them was the novel thought that humans were not at the centre of things. The Christian Church viewed this suggestion as contrary to doctrine and did its best to discourage it. Explicit heliocentrism was heresy.

It prevailed nevertheless, because the evidence was so strong. New and better heliocentric theories appeared. Then the spheres were thrown away altogether, in favour of a different shape from classical geometry: the ellipse. Ellipses are oval shapes, and indirect evidence suggests they were first studied in Greek geometry by Menaechmus around 350 BC, along with hyperbolas and parabolas, as sections of a cone, Figure 13. Euclid is said to have written four books on conic sections, though nothing has survived if he did, and Archimedes investigated some of their properties. Greek research on the topic reached its climax in about 240 BC with the eight-volume *Conic Sections* by Apollonius of Perga, who found a way to define

ellipse   parabola   hyperbola

Fig 13 Conic sections.

these curves purely within a plane, avoiding the third dimension. However, the Pythagorean view that circles and spheres attained a higher degree of perfection than ellipses and other more complex curves persisted.

Ellipses cemented their role in astronomy around 1600, with the work of Kepler. His astronomical interests began in childhood; at the age of six he witnessed the great comet of 1577,[2] and three years later he saw an eclipse of the Moon. At the University of Tübingen, Kepler showed great talent for mathematics and put it to profitable use casting horoscopes. In those days mathematics, astronomy, and astrology often went together. He combined a heady level of mysticism with a level-headed attention to mathematical detail. A typical example is his *Mysterium Cosmographicum* ('The Cosmographic Mystery'), a spirited defence of the heliocentric system published in 1596. It combines a clear grasp of Copernicus's theory with what to modern eyes is a very strange speculation relating the distances of the known planets from the Sun to the regular solids. For a long time Kepler regarded this discovery as one of his greatest, revealing the Creator's plan for the universe. He saw his later researches, which we now consider to be far more significant, as mere elaborations of this basic plan. At the time, one advantage of the theory was that it explained why there were precisely six planets (Mercury through Saturn). Between these six orbits lie five gaps, one for each regular solid. With the discovery of Uranus and later Neptune and Pluto (until its recent demotion from planetary status) this feature quickly became a fatal flaw.

Kepler's lasting contribution has its roots in his employment by Tycho Brahe. The two first met in 1600. After a two-month stay and a heated argument Kepler negotiated an acceptable salary. Following a spate of problems in his home city of Graz he moved to Prague, assisting Tycho in the analysis of his planetary observations, especially of Mars. When Tycho unexpectedly died in 1601 Kepler took over his employer's position as imperial mathematician to Rudolph II. His primary role was casting imperial horoscopes, but he also had time to continue his analysis of the orbit of Mars. Following traditional epicyclic principles he refined his model to the point at which its errors, compared with observation, were usually a mere two minutes of arc, the typical error in the observations themselves. However, he didn't stop there because sometimes the errors were bigger, up to eight minutes of arc.

His search eventually led him to two laws of planetary motion, published in *Astronomia Nova* ('A New Astronomy'). For many years he had tried to fit the orbit of Mars to an ovoid – an egg-shaped curve, sharper at one end than the other – without success. Perhaps he expected the orbit to be more curved closer to the Sun. In 1605 it occurred to Kepler to try an ellipse, equally rounded at both ends, and to his surprise this did a much better job. He concluded that all planetary orbits are ellipses, his first law. His second law described how the planet moves along its orbit, stating that planets sweep out equal areas in equal times. The book appeared in 1609. Kepler then devoted much of his effort to preparing various astronomical tables, but he returned to the regularities of planetary orbits in 1619 in his *Harmonices Mundi* ('The Harmony of the World'). This book had some ideas we now find strange, for example that the planets emit musical sounds as they roll round the Sun. But it also includes his third law: the squares of the orbital periods are proportional to the cubes of the distances from the Sun.

Kepler's three laws were all but buried amid a mass of mysticism, religious symbolism, and philosophical speculation. But they represented a giant leap forward, leading Newton to one of the greatest scientific discoveries of all time.

Newton derived his law of gravity from Kepler's three laws of planetary motion. It states that every particle in the universe attracts every other particle with a force that is proportional to the product of their masses and inversely proportional to the square of the distance between them. In symbols,

$$F = G\frac{m_1 m_2}{d^2}$$

Here F is the attractive force, d is the distance, the ms are the two masses, and G is a specific number, the gravitational constant.[3]

Who discovered Newton's law of gravity? It sounds like one of those self-answering questions, like 'whose statue stands on top of Nelson's column?'. But a reasonable answer is the curator of experiments at the Royal Society, Robert Hooke. When Newton published the law in 1687, in his *Principia*, Hooke accused him of plagiarism. However, Newton provided the first mathematical derivation of elliptical orbits from the law, which was vital in establishing its correctness, and Hooke acknowledged this. Moreover, Newton had cited Hooke, along with several others, in the book. Presumably Hooke felt he deserved more credit; he had suffered similar problems several times before and it was a sore point.

The idea that bodies attract each other had been floating around for a while, and so had its likely mathematical expression. In 1645 the French astronomer Ismaël Boulliau (Bullialdus) wrote his *Astronomia Philolaica*

('Philolaic Astronomy' – Philolaus was a Greek philosopher who thought that a central fire, not the Earth, was the centre of the universe). In it he wrote:

> As for the power by which the Sun seizes or holds the planets, and which, being corporeal, functions in the manner of hands, it is emitted in straight lines throughout the whole extent of the world, and like the species of the Sun, it turns with the body of the Sun; now, seeing that it is corporeal, it becomes weaker and attenuated at a greater distance or interval, and the ratio of its decrease in strength is the same as in the case of light, namely, the duplicate proportion, but inversely, of the distances.

This is the famous 'inverse square' dependency of the force on distance. There are simple, though naive, reasons to expect such a formula, because the surface area of a sphere varies as the square of its radius. If the same amount of gravitational 'stuff' spreads out over ever-increasing spheres as it departs from the Sun, then the amount of it received at any point must vary in the inverse proportion to the surface area. Exactly this happens with light, and Boulliau assumed, without much evidence, that gravity must be analogous. He also thought that the planets move along their orbits under their own power, so to speak: 'No kind of motion presses upon the remaining planets, [which] are driven round by individual forms with which they were provided.'

Hooke's contribution dates to 1666, when he presented a paper to the Royal Society with the title 'On gravity'. Here he sorted out what Boulliau had got wrong, arguing that an attractive force from the Sun could interfere with a planet's natural tendency to move in a straight line (as specified by Newton's third law of motion) and cause it to follow a curve. He also stated that 'these attractive powers are so much the more powerful in operating, by how much the nearer the body wrought upon is to their own Centers', showing that he thought the force fell off with distance. But he didn't tell anyone else the mathematical form for this decrease until 1679, when he wrote to Newton: 'The Attraction always is in a duplicate proportion to the Distance from the Center Reciprocall.' In the same letter he said that this implies that the velocity of a planet varies as the reciprocal of its distance from the Sun. Which is wrong.

When Hooke complained that Newton had stolen his law, Newton was having none of it, pointing out that he had discussed the idea with Christopher Wren before Hooke had sent his letter. To demonstrate prior art, he cited Boulliau, and also Giovanni Borelli, an Italian physiologist and mathematical physicist. Borelli had suggested that three forces combine to create planetary motion: an inward force caused by the planet's desire to approach the Sun, a sideways force caused by sunlight, and an outward force caused by the Sun's rotation. Score one out of three, and that's generous.

Newton's main point, generally considered decisive, is that whatever else Hooke had done, he had not deduced the exact form of orbits from inverse square law attraction. Newton had. In fact, he had deduced all three of Kepler's laws of planetary motion: elliptical orbits, sweeping out equal areas in equal intervals of time, with the square of the period being proportional to the cube of the distance. 'Without my Demonstrations,' Newton insisted, the inverse square law 'cannot be believed by a judicious philosopher to be anywhere accurate.' But he did also accept that 'Mr Hook is yet a stranger' to this proof. A key feature of Newton's argument is that it applies not just to a point particle, but to a sphere. This extension, which is crucial to planetary motion, had caused Newton considerable effort. His geometric proof is a disguised application of integral calculus, and he was justifiably proud of it. There is also documentary evidence that Newton had been thinking about such questions for quite a while.

At any rate, we name the law after Newton, and this does justice to the importance of his contribution.

The most important aspect of Newton's law of gravitation is not the inverse square law as such. It is the assertion that gravitation acts universally. *Any* two bodies, anywhere in the universe, attract each other. Of course you need an accurate force law (inverse square) to get accurate results, but without universality, you don't know how to write down the equations for any system with more than two bodies. Almost all of the interesting systems, such as the Solar System itself, or the fine structure of the motion of the Moon under the influence of (at least) the Sun and the Earth, involve more than two bodies, so Newton's law would have been almost useless if it had applied only to the context in which he first deduced it.

What motivated this vision of universality? In his 1752 *Memoirs of Sir Isaac Newton's Life*, William Stukeley reported a tale Newton had told him in 1726:

> The notion of gravitation ... was occasioned by the fall of an apple, as he sat in contemplative mood. Why should that apple always descend

perpendicularly to the ground, thought he to himself. Why should it not go sideways or upwards, but constantly to the Earth's centre? Assuredly the reason is, that the Earth draws it. There must be a drawing power in matter. And the sum of the drawing power in the matter of the Earth must be in the Earth's centre, not in any side of the Earth. Therefore does this apple fall perpendicularly or towards the centre? If matter thus draws matter; it must be in proportion of its quantity. Therefore the apple draws the Earth, as well as the Earth draws the apple.

Whether the story is the literal truth, or a convenient fiction that Newton invented to help him explain his ideas later on, is not entirely clear, but it seems reasonable to take the tale at face value because the idea does not end with apples. The apple was important to Newton because it made him realise that the same law of forces can explain both the motion of the apple and that of the Moon. The only difference is that the Moon also moves sideways; this is why it stays up. Actually, it is always falling towards the Earth, but the sideways motion causes the Earth's surface to fall away as well. Newton, being Newton, didn't stop with this qualitative argument. He did the sums, compared them with observations, and was satisfied that his idea must be correct.

If gravity acts on the apple, the Moon, and the Earth, as an inherent feature of matter, then presumably it acts on everything.

It is not possible to verify the universality of gravitational forces directly; you would have to study all pairs of bodies in the entire universe, and find a way to remove the influence of all the other bodies. But that's not how science works. Instead, it employs a mixture of inference and observations. Universality is a hypothesis, capable of being falsified every time it is applied. Every time it survives falsification – a fancy way to say it gives good results – the justification for using it becomes a little stronger. If (as in this case) it survives thousands of such tests, the justification becomes very strong indeed. However, the hypothesis can never be proved *true*: for all we know, the next experiment might produce incompatible results. Perhaps somewhere in a galaxy far, far away there is one speck of matter, one atom, that is not attracted to everything else. If so, we will never find it; equally, it won't upset our calculations. The inverse square law itself is exceedingly difficult to verify directly, that is, by actually measuring the attractive force. Instead, we apply the law to systems that we can measure by using it to predict orbits, and then check whether the predictions agree with observations.

Even granting universality, it is not enough to write down an accurate law of attraction. That just produces an equation describing the motion. In order to find the motion itself, you have to solve the equation. Even for two bodies, this is not straightforward, and even bearing in mind that he knew in advance what answer to expect, Newton's deduction of elliptical orbits is a *tour de force*. It explains why Kepler's three laws provide a very accurate description of each planet's orbit. It also explains why that description is not exact: other bodies in the solar system, other than the Sun and the planet itself, affect the motion. In order to account for these disturbances, you have to solve the equations of motion for three or more bodies. In particular, if you want to predict the motion of the Moon with high precision, you have to include the Sun and the Earth in your equations. The effects of the other planets, especially Jupiter, are not entirely negligible either, but they show up only in the long term. So, fresh from Newton's success with the motion of two bodies under gravity, mathematicians and physicists moved on to the next case: three bodies. Their initial optimism dissipated rapidly: the three-body case turned out to be very different from the two-body case. In fact, it defied solution.

It was often possible to calculate good *approximations* to the motion (which often solved the problem for practical purposes), but there no longer seemed to be an exact formula. This problem bedevilled even simplified versions, such as the restricted three-body problem. Suppose that a planet orbits a star in a perfect circle: how will a speck of dust, of negligible mass, move?

Calculating approximate orbits for three or more bodies, by hand, using pencil and paper, was just about feasible, but very laborious. Mathematicians devised innumerable tricks and short cuts, leading to a reasonable understanding of several astronomical phenomena. Only in the late nineteenth century did the true complexity of the three-body problem become apparent, when Henri Poincaré realised that the geometry involved was necessarily extraordinarily intricate. And only in the late twentieth century did the advent of powerful computers reduce the labour of hand calculations, permitting accurate long-term predictions of the motion of the Solar System.

Poincaré's breakthrough – if it can be called that, since at the time it seemed to be telling everyone that the problem was hopeless and it was pointless to seek a solution – came about because he competed for a mathematical prize. Oscar II, king of Sweden and Norway, announced a competition to celebrate his 60th birthday in 1889. Taking advice from the mathematician Gösta Mittag-Leffler, the king chose the general problem of arbitrarily many bodies moving under Newtonian gravitation. Since it was well understood that an explicit formula akin to the two-body ellipse was an unrealistic aim, the requirement was relaxed: the prize would be awarded for an approximation method of a very specific kind. Namely, the motion must be determined as an infinite series, giving results as accurate as we please if enough terms are included.

Poincaré did not answer this question. Instead, his memoir on the topic, published in 1890, provided evidence that it might not possess that kind of answer, even for just three bodies – star, planet, and dust particle. By thinking about the geometry of hypothetical solutions, Poincaré discovered that in some cases the orbit of the dust particle must be exceedingly complex and tangled. He then, in effect, threw up his hands in horror and made the pessimistic statement that 'When one tries to depict the figure formed by these two curves and their infinity of intersections, each of which corresponds to a doubly asymptotic solution, these intersections form a kind of net, web or infinitely tight mesh... One is struck by the complexity of this figure that I am not even attempting to draw.'

We now see Poincaré's work as a breakthrough, and discount his pessimism, because the complicated geometry that led him to despair of ever solving the problem actually provides powerful insights if it is properly developed and understood. The complex geometry of the associated dynamics turned out to be one of the earliest examples of *chaos*: the occurrence, in non-random equations, of solutions so complicated that in some respects they appear to be random, see Chapter 16.

There are several ironies in the story. Mathematical historian June Barrow-Green discovered that the published version of Poincaré's prizewinning memoir was not the one that won the prize.[4] This earlier version contained a major error, overlooking the chaotic solutions. The work was at proof stage when an embarrassed Poincaré realised his blunder, and he paid for a new printing of a corrected version. Almost all copies of the original were destroyed, but one remained tucked away in the archives of the Mittag-Leffler Institute in Sweden, where Barrow-Green found it.

It also turned out that the presence of chaos does not, in fact, rule out series solutions, but these are valid almost always rather than always. Karl Frithiof Sundman, a Finnish mathematician, discovered this in 1912 for the three-body problem, using series formed from powers of the cube root of time. (Powers of time won't hack it.) The series converge – have a sensible sum – unless the initial state has zero angular momentum, but such states are infinitely rare, in the sense that a random choice of angular momentum is almost always nonzero. In 1991 the Chinese mathematician Qiudong Wang extended these results to any number of bodies, but did not classify the rare exceptions when the series fail to converge. Such a classification is likely to be very complicated: it must include solutions where bodies escape to infinity in finite time, or oscillate ever faster, both of which can happen for five or more bodies.

Newton's law of gravity is routinely applied to design orbits for space missions. Here even two-body dynamics is useful in its own right. In its early days, the exploration of the Solar System mainly used two-body orbits, segments of ellipses. By burning its rockets the spacecraft could be switched from one ellipse to a different one. But as the aims of space programmes got more ambitious, more efficient methods were needed. They came from many-body dynamics, usually three bodies but occasionally as great as five. The new methods of chaos and topological dynamics became the basis of practical solutions to engineering problems.

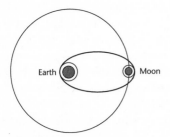

Fig 14 Hohmann transfer ellipse from low-Earth orbit to lunar orbit.

It all started with a simple question: What is the most efficient route from the Earth to the Moon or the planets? The classic answer, known as a Hohmann transfer ellipse (Figure 14), starts from a circular orbit round the Earth, and then follows part of a long, thin ellipse to join up with a second circular orbit round the destination. This method was employed for the Apollo missions of the 1960s and 1970s, but for many types of mission it has one disadvantage. The spacecraft must be boosted out of Earth orbit and slowed again to enter lunar orbit; this wastes fuel. There are alternatives involving many loops round the Earth, a transition through the point between Earth and Moon where their gravitational fields cancel, and many loops round the Moon. But trajectories like that take longer than Hohmann ellipses, so they were not used for the manned Apollo missions where food and oxygen, hence time, were of the essence. For unmanned missions, however, time is relatively cheap, whereas anything that adds to the overall weight of the spacecraft, including fuel, costs money.

By taking a fresh look at Newton's law of gravity and his second law of motion, mathematicians and space engineers have recently discovered a new, and remarkable, approach to fuel-efficient interplanetary travel.

Go by tube.

It's an idea straight out of science fiction. In his 2004 *Pandora's Star*, Peter Hamilton portrays a future where people travel to planets encircling distant stars by train, running the railway lines through a wormhole, a short cut through space-time. In his Lensman series from 1934 to 1948, Edward Elmer 'Doc' Smith came up with the hyperspatial tube, which malevolent aliens used to invade human worlds from the fourth dimension.

Although we don't yet have wormholes or aliens from the fourth dimension, it has been discovered that the planets and moons of the Solar System are tied together by a network of tubes, whose mathematical definition requires many more dimensions than four. The tubes provide energy-efficient routes from one world to another. They can be seen only through mathematical eyes, because they are not made of matter: their walls are energy levels. If we could visualise the ever-changing landscape of gravitational fields that controls how the planets move, we would be able to see the tubes, swirling along with the planets as they orbit the Sun.

Tubes explain some puzzling orbital dynamics. Consider, for example, the comet called Oterma. A century ago, Oterma's orbit was well outside that of Jupiter. But after a close encounter with the giant planet, the comet's orbit shifted inside that of Jupiter. After another close encounter, it switched back outside again. We can confidently predict that Oterma will continue to switch orbits in this way every few decades: not because it breaks Newton's law, but because it obeys it.

This is a far cry from tidy ellipses. The orbits predicted by Newtonian gravity are elliptical only when no other bodies exert a significant gravitational pull. But the Solar System is full of other bodies, and they can make a huge – and surprising – difference. It is here that the tubes enter the story. Oterma's orbit lies inside two tubes, which meet near Jupiter. One tube lies inside Jupiter's orbit, the other outside. They enclose special orbits in 3 : 2 and 2 : 3 resonance with Jupiter, meaning that a body in such an orbit will go round the Sun three times for every two revolutions of Jupiter, or two times for every three. At the tube junction near Jupiter, the comet can switch tubes, or not, depending on rather subtle effects of Jovian and solar gravity. But once inside a tube, Oterma is stuck there until the tube returns to the junction. Like a train that has to stay on the rails, but can change its route to another set of rails if someone switches the points, Oterma has some freedom to change its itinerary, but not a lot (Figure 15).

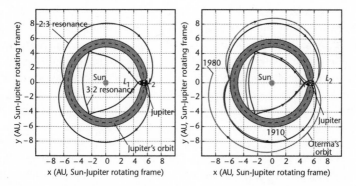

Fig 15 *Left*: Two periodic orbits, in 2 : 3 and 3 : 2 resonance with Jupiter, connected via Lagrange points. *Right*: Actual orbit of comet Oterma, 1910–1980.

The tubes and their junctions may seem bizarre, but they are natural and important features of the gravitational geography of the Solar System. Victorian railway-builders understood the need to exploit natural features of the landscape, running railways through valleys and along contour lines, and digging tunnels through hills rather than taking the train over the top. One reason was that trains tend to slip on steep gradients, but the main one was energy. Climbing a hill, against the force of gravity, costs energy, which shows up as increased fuel consumption, which costs money.

It's much the same with interplanetary travel. Imagine a spacecraft moving through space. Where it goes next does not depend solely on where it is now: it also depends on how fast it is moving and in which direction. It takes three numbers to specify the spacecraft's position – for example its direction from the Earth, which requires two numbers (astronomers use right ascension and declination, which are analogous to longitude and latitude on the celestial sphere, the apparent sphere formed by the night sky), and its distance from the Earth. It takes a further three numbers to specify its velocity in those three directions. So the spacecraft travels through a mathematical landscape that has six dimensions rather than two.

A natural landscape is not flat: it has hills and valleys. It takes energy to climb a hill, but a train can gain energy by rolling down into a valley. In fact, two types of energy come into play. The height above sea-level determines the train's potential energy, which represents work done against the force of gravity. The higher you go, the more potential energy you must create. The second kind is kinetic energy, which corresponds to speed. The faster you go, the greater your kinetic energy becomes. When the train rolls downhill and accelerates, it trades potential energy for kinetic. When it climbs a hill and slows down, the trade is in the reverse direction. The total energy is constant, so the train's trajectory is analogous to a contour line in the energy landscape. However, trains have a third source of energy: coal, diesel, or electricity. By expending fuel, a train can climb a gradient or speed up, freeing itself from its natural free-running trajectory. The total energy still cannot change, but all else is negotiable.

It is much the same with spacecraft. The combined gravitational fields of the Sun, planets, and other bodies of the Solar System provide potential energy. The speed of the spacecraft corresponds to kinetic energy. And its motive power – be it rocket fuel, ions, or light-pressure – adds a further energy source, which can be switched on or off as required. The path followed by the spacecraft is a kind of contour line in the corresponding energy landscape, and along that path the total energy remains constant. And some types of contour line are surrounded by tubes, corresponding to nearby energy levels.

Those Victorian railway engineers were also aware that the terrestrial landscape has special features – peaks, valleys, mountain passes – which have a big effect on efficient routes for railway lines, because they constitute a kind of skeleton for the overall geometry of the contours. For instance, near a peak or a valley bottom the contours form closed curves. At peaks, potential energy is locally at a maximum; in a valley, it is at a local minimum. Passes combine features of both, being at a maximum in one direction, but a minimum in another. Similarly, the energy landscape of the Solar System has special features. The most obvious are the planets and moons themselves, which sit at the bottom of gravity wells, like valleys. Equally important, but less visible, are the peaks and passes of the energy landscape. All these features organise the overall geometry, and with it, the tubes.

The energy landscape has other attractive features for the tourist, notably Lagrange points. Imagine a system consisting only of the Earth and the Moon. In 1772 Joseph-Louis Lagrange discovered that at any instant there are precisely five places where the gravitational fields of the two bodies, together with centrifugal force, cancel out exactly. Three are in line with both Earth and Moon – L1 lies between them, L2 is on the far side of the Moon, and L3 is on the far side of the Earth. The Swiss mathematician Leonhard Euler had already discovered these around 1750. But there are also L4 and L5, known as Trojan points, which lie in the same orbit as the Moon but 60 degrees ahead of it or behind it. As the Moon rotates round the Earth, the Lagrange points rotate with it. Other pairs of bodies also have Lagrange points – Earth/Sun, Jupiter/Sun, Titan/Saturn.

The old-fashioned Hohmann transfer orbit is built from pieces of circles and ellipses, which are the natural trajectories for two-body systems. The new tube-based paths are built from pieces of the natural trajectories of three-body systems, such as Sun/Earth/spacecraft. Lagrange points play a special role, just as peaks and passes did for railways: they are the junctions where tubes meet. L1 is a great place to make small course changes, because the natural dynamics of a spacecraft near L1 is chaotic, Figure 16. Chaos has a useful feature (see Chapter 16): very small changes in position or speed can create large changes to the trajectory. So it is easy to redirect the spacecraft in a fuel-efficient, though possibly slow, manner.

The first person to take this idea seriously was the German-born mathematician Edward Belbruno, an orbital analyst at the Jet Propulsion Laboratory from 1985 to 1990. He realised that chaotic dynamics in many-body systems provided an opportunity for novel low-energy transfer orbits, naming the technique fuzzy boundary theory. In 1991 he put his ideas into practice. Hiten, a Japanese probe, had been surveying the Moon, and had completed its intended mission, returning to orbit the Earth. Belbruno designed a new orbit that would take it back to the Moon despite having pretty much run out of fuel. After approaching the Moon as intended, Hiten visited its L4 and L5 points to search for cosmic dust that might have been trapped there.

A similar trick was used in 1985 to redirect the almost-dead International Sun–Earth Explorer ISEE-3 to rendezvous with comet Giacobini–Zinner, and it was used again for NASA's Genesis mission to bring back samples of the solar wind. Mathematicians and engineers

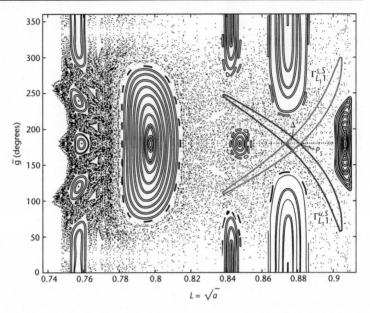

Fig 16 Chaos near Jupiter. The diagram shows a cross-section of orbits. The nested loops are quasiperiodic orbits and the remaining stippled region is a chaotic orbit. The two thin loops crossing each other at the right are cross-sections of tubes.

wanted to repeat the trick, and to find others of the same kind, which meant finding out what really made it work. It turned out to be tubes.

The underlying idea is simple but clever. Those special places in the energy landscape that resemble mountain passes create bottlenecks that would-be travellers cannot easily avoid. Ancient humans discovered, the hard way, that even though it takes energy to climb a pass, it takes *more* energy to follow any other route – unless you can go round the mountain in a totally different direction. The pass makes the best of a bad choice.

In the energy landscape, the analogues of passes include Lagrange points. Associated with them are very specific inbound paths, which are like the most efficient way to climb up the pass. There are also equally specific outbound paths, analogous to the natural routes down from the pass. To follow these inbound and outbound paths exactly, you have to travel at just the right speed, but if your speed is slightly different you can still stay near those paths. In the late 1960s American mathematicians Charles Conley and Richard McGehee followed up Belbruno's pioneering work, pointing out that each such path is surrounded by a nested set of tubes, one inside the other. Each tube corresponds to a particular choice of speed; the further away it is from the optimal speed, the wider the tube is.

On the surface of any given tube, the total energy is constant, but the constants differ from one tube to another. Much as a contour line is at a constant height, but that height is different for each contour.

The way to plan an efficient mission profile, then, is to work out which tubes are relevant to your choice of destination. Then you route your spacecraft along the inside of the first inbound tube, and when it gets to the associated Lagrange point you fire a quick burst on the motors to redirect it along the most suitable outbound tube, Figure 17. That tube naturally flows into the corresponding inbound tube of the next switching point... and so it goes.

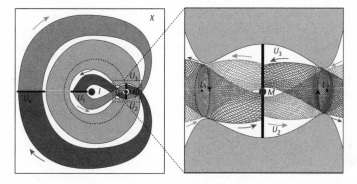

Fig 17 *Left*: Tubes meeting near Jupiter. *Right*: Close-up of region where the tubes join.

Plans for future tubular missions are already being drawn up. In 2000 Wang Sang Koon, Martin Lo, Jerrold Marsden, and Shane Ross used the tube technique to find a 'Petit Grand Tour' of the moons of Jupiter, ending with a capture orbit round Europa, which was very tricky with previous methods. The path involves a gravitational boost near Ganymede followed by a tube trip to Europa. A more complex route, requiring even less energy, includes Callisto as well. It makes use of another feature of the energy landscape – resonances. These occur when, say, two moons repeatedly return to the same relative positions, but one revolves twice round Jupiter while the other revolves three times. Any small numbers can replace 2 and 3 here. This route uses five-body dynamics: Jupiter, the three moons, and the spacecraft.

In 2005, Michael Dellnitz, Oliver Junge, Marcus Post, and Bianca Thiere used tubes to plan an energy-efficient mission from the Earth to Venus. The main tube here links the Sun/Earth L1 point to the Sun/Venus L2 point. As a comparison, this route uses only one third of the fuel required by the European Space Agency's Venus Express mission, because it can use low-thrust engines; the price paid is a lengthening of the transit time from 150 days to about 650 days.

The influence of tubes may go further. In unpublished work, Dellnitz has discovered evidence of a natural system of tubes connecting Jupiter to each of the inner planets. This remarkable structure, now called the Interplanetary Superhighway, hints that Jupiter, long known to be the dominant planet of the Solar System, also plays the role of a celestial Grand Central Station. Its tubes may well have organised the formation of the entire Solar System, determining the spacings of the inner planets.

Why were the tubes not spotted sooner? Until very recently, two vital things were missing. One was powerful computers, capable of carrying out the necessary many-body calculations. They are far too cumbersome by hand. But the other, even more important, was a deep mathematical understanding of the geography of the energy landscape. Without this imaginative triumph of modern mathematical methods, there would be nothing for the computers to calculate. And without Newton's law of gravity, the mathematical methods would never have been devised.

5 Portent of the ideal world
The Square Root of Minus One

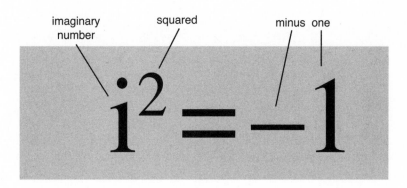

imaginary number

squared

minus one

$$i^2 = -1$$

What does it say?

Even though it ought to be impossible, the square of the number i is minus one.

Why is that important?

It led to the creation of complex numbers, which in turn led to complex analysis, one of the most powerful areas of mathematics.

What did it lead to?

Improved methods to calculate trigonometric tables. Generalisations of almost all mathematics to the complex realm. More powerful methods to understand waves, heat, electricity, and magnetism. The mathematical basis of quantum mechanics.

Renaissance Italy was a hotbed of politics and violence. The north of the country was controlled by a dozen warring city-states, among them Milan, Florence, Pisa, Genoa, and Venice. In the south, Guelphs and Gibellines were in conflict as Popes and Holy Roman Emperors battled for supremacy. Bands of mercenaries roamed the land, villages were laid waste, coastal cities waged naval warfare against each other. In 1454 Milan, Naples, and Florence signed the Treaty of Lodi, and peace reigned for the next four decades, but the papacy remained embroiled in corrupt politics. This was the time of the Borgias, notorious for poisoning anyone who got in the way of their quest for political and religious power, but it was also the time of Leonardo da Vinci, Brunelleschi, Piero della Francesca, Titian, and Tintoretto. Against a backdrop of intrigue and murder, long-held assumptions were coming into question. Great art and great science flourished in symbiosis, each feeding off the other.

Great mathematics flourished as well. In 1545 the gambling scholar Girolamo Cardano was writing an algebra text, and he encountered a new kind of number, one so baffling that he declared it 'as subtle as it is useless' and dismissed the notion. Rafael Bombelli had a solid grasp of Cardano's algebra book, but he found the exposition confusing, and decided he could do better. By 1572 he had noticed something intriguing: although these baffling new numbers made no sense, they could be used in algebraic calculations and led to results that were demonstrably correct.

For centuries mathematicians engaged in a love–hate relationship with these 'imaginary numbers', as they are still called today. The name betrays an ambivalent attitude: they're not *real* numbers, the usual numbers encountered in arithmetic, but in most respects they behave like them. The main difference is that when you square an imaginary number, the result is negative. But that ought not to be possible, because squares are always positive.

Only in the eighteenth century did mathematicians figure out what imaginary numbers were. Only in the nineteenth did they start to feel comfortable with them. But by the time the logical status of imaginary numbers was seen to be entirely comparable to that of the more traditional real numbers, imaginaries had become indispensable throughout mathematics and science, and the question of their meaning hardly seemed interesting any more. In the late nineteenth and early twentieth centuries, revived interest in the foundations of mathematics led to a rethink of the concept of number, and traditional 'real' numbers were seen to be no more real than imaginary ones. Logically, the two kinds of number were as alike as Tweedledum and Tweedledee. Both were constructs of the human mind, both represented – but were not synonymous with – aspects of nature. But they represented reality in different ways and in different contexts.

By the second half of the twentieth century, imaginary numbers were simply part and parcel of every mathematician's and every scientist's mental toolkit. They were built into quantum mechanics in such a fundamental way that you could no more do physics without them than you could scale the north face of the Eiger without ropes. Even so, imaginary numbers are seldom taught in schools. The sums are easy enough, but the mental sophistication needed to appreciate why imaginaries are worth studying is still too great for the vast majority of students. Very few adults, even educated ones, are aware of how deeply their society depends on numbers that do not represent quantities, lengths, areas, or amounts of money. Yet most modern technology, from electric lighting to digital cameras, could not have been invented without them.

Let me backtrack to a crucial question. *Why* are squares always positive?

In Renaissance times, where equations were generally rearranged to make every number in them positive, they wouldn't have phrased the question quite this way. They would have said that if you add a number to a square then you have to get a bigger number – you can't get zero. But even if you allow negative numbers, as we now do, squares still have to be positive. Here's why.

Real numbers can be positive or negative. However, the square of any real number, whatever its sign, is always positive, because the product of two negative numbers is positive. So both 3×3 and -3×-3 yield the same result: 9. Therefore 9 has *two* square roots, 3 and -3.

What about -9? What are its square roots?

It doesn't have any.

It all seems terribly unfair: the positive numbers hog two square roots each, while the negative numbers go without. It is tempting to change the rule for multiplying two negative numbers, so that, say, $-3 \times -3 = -9$. Then positive and negative numbers each get one square root; moreover, this has the same sign as its square, which seems neat and tidy. But this seductive line of reasoning has an unintended downside: it wrecks the usual rules of arithmetic. The problem is that -9 already occurs as 3×-3 itself a consequence of the usual rules of arithmetic, and a fact that almost everyone is happy to accept. If we insist that -3×-3 is also -9, then $-3 \times -3 = 3 \times -3$. There are several ways to see that this causes problems; the simplest is to divide both sides by -3, to get $3 = -3$.

Of course you can change the rules of arithmetic. But now it all gets complicated and messy. A more creative solution is to retain the rules of arithmetic, and to extend the system of real numbers by permitting imaginaries. Remarkably – and no one could have anticipated this, you just have to follow the logic through – this bold step leads to a beautiful, consistent system of numbers, with a myriad uses. Now all numbers except 0 have *two* square roots, one being minus the other. This is true even for the new kinds of number; one enlargement of the system suffices. It took a while for this to become clear, but in retrospect it has an air of inevitability. Imaginary numbers, impossible though they were, refused to go away. They seemed to make no sense, but they kept cropping up in calculations. Sometimes the use of imaginary numbers made the calculations simpler, and the result was more comprehensive and more satisfactory. Whenever an answer that had been obtained using imaginary numbers, but did not explicitly involve them, could be verified independently, it turned out to be right. But when the answer did involve explicit imaginary numbers it seemed to be meaningless, and often logically contradictory. The enigma simmered for two hundred years, and when it finally boiled over, the results were explosive.

Cardano is known as the gambling scholar because both activities played a prominent role in his life. He was both genius and rogue. His life consists of a bewildering series of very high highs and very low lows. His mother tried to abort him, his son was beheaded for killing his (the son's) wife, and he (Cardano) gambled away the family fortune. He was accused of heresy for casting the horoscope of Jesus. Yet in between he also became Rector of the University of Padua, was elected to the College of Physicians in Milan, gained 2000 gold crowns for curing the Archbishop of St Andrews' asthma, and received a pension from Pope Gregory XIII. He invented the combination lock and gimbals to hold a gyroscope, and he wrote a number of books, including an extraordinary autobiography *De Vita Propria* ('The Book of My Life'). The book that is relevant to our tale is the *Ars Magna* of 1545. The title means 'great art', and refers to algebra. In it, Cardano assembled the most advanced algebraic ideas of his day, including new and dramatic methods for solving equations, some invented by a student of his, some obtained from others in controversial circumstances.

Algebra, in its familiar sense from school mathematics, is a system for representing numbers symbolically. Its roots go back to the Greek Diophantus around 250 AD, whose *Arithmetica* employed symbols to describe ways to solve equations. Most of the work was verbal – 'find two numbers whose sum is 10 and whose product is 24'. But Diophantus summarised the methods he used to find the solutions (here 4 and 6) symbolically. The symbols (see Table 1) were very different from those we use today, and most were abbreviations, but it was a start. Cardano mainly used words, with a few symbols for roots, and again the symbols scarcely resemble those in current use. Later authors homed in, rather haphazardly, on today's notation, most of which was standardised by Euler in his numerous textbooks. However, Gauss still used xx instead of x^2 as late as 1800.

date	author	notation
c.250	Diophantus	$\Delta^Y a\varsigma\beta\overset{\circ}{M}\gamma$
c.825	Al-Khowârizmî	*power plus twice side plus three* [in Arabic]
1545	Cardano	*square plus twice side plus three* [in Italian]
1572	Bombelli	$3p \cdot 2 \overset{1}{\smile} p \cdot 1 \overset{2}{\smile}$
1585	Stevin	$3 + 2^{①} + 1^{②}$
1591	Viète	x quadr. $+\, x\, 2 + 3$
1637	Descartes, Gauss	$xx + 2x + 3$
1670	Bachet de Méziriac	$Q + 2N + 3$
1765	Euler, modern	$x^2 + 2x + 3$

Table 1 The development of algebraic notation.

The most important topics in the *Ars Magna* were new methods for solving cubic and quartic equations. These are like quadratic equations, which most of us meet in school algebra, but more complicated. A quadratic equation states a relationship involving an unknown quantity, normally symbolised by the letter x, and its square x^2. 'Quadratic' comes from the Latin for 'square'. A typical example is

$$x^2 - 5x + 6 = 0$$

Verbally, this says: 'Square the unknown, subtract 5 times the unknown, and add 6: the result is zero.' Given an equation involving an unknown, our task is to solve the equation – to find the value or values of the unknown that make the equation correct.

For a randomly chosen value of x, this equation will usually be false. For example, if we try $x = 1$, then $x^2 - 5x + 6 = 1 - 5 + 6 = 2$, which isn't zero. But for rare choices of x, the equation is true. For example, when $x = 2$ we have $x^2 - 5x + 6 = 4 - 10 + 6 = 0$. But this is not the *only* solution! When $x = 3$ we have $x^2 - 5x + 6 = 9 - 15 + 6 = 0$ as well. There are two solutions, $x = 2$ and $x = 3$, and it can be shown that there are no others. A quadratic equation can have two solutions, one, or none (in real numbers). For example, $x^2 - 2x + 1 = 0$ has only the solution $x = 1$, and $x^2 + 1 = 0$ has no solutions in real numbers.

Cardano's masterwork provides methods for solving cubic equations, which along with x and x^2 also involve the cube x^3 of the unknown, and quartic equations, where x^4 turns up as well. The algebra gets very complicated; even with modern symbolism it takes a page or two to derive the answers. Cardano did not go on to quintic equations, involving x^5, because he did not know how to solve them. Much later it was proved that no solutions (of the type Cardano would have wanted) exist: although highly accurate numerical solutions can be calculated in any particular case, there is no general *formula* for them, unless you invent new symbols specifically for the task.

I'm going to write down a few algebraic formulas, because I think the topic makes more sense if we don't try to avoid them. You don't need to follow the details, but I'd like to show you what everything looks like. Using modern symbols, we can write out Cardano's solution of the cubic equation in a special case, when $x^3 + ax + b = 0$ for specific numbers a and b. (If x^2 is present, a cunning trick gets rid of it, so this case actually deals with everything.) The answer is:

$$x = \sqrt[3]{-\frac{b}{2} + \sqrt{\frac{b^2}{4} + \frac{a^3}{27}}} + \sqrt[3]{-\frac{b}{2} - \sqrt{\frac{b^2}{4} + \frac{a^3}{27}}}$$

This may appear a bit of a mouthful, but it's a lot simpler than many algebraic formulas. It tells us how to calculate the unknown x by working out the square of b and the cube of a, adding a few fractions, and taking a couple of square roots (the $\sqrt{}$ symbol) and a couple of cube roots (the $\sqrt[3]{}$ symbol). The cube root of a number is whatever you have to cube to get that number.

The discovery of the solution for cubic equations involves at least three other mathematicians, one of whom complained bitterly that Cardano had promised not to reveal his secret. The story, though fascinating, is too complicated to relate here.[1] The quartic was solved by Cardano's student Lodovico Ferrari. I'll spare you the even more complicated formula for quartic equations.

The results reported in the *Ars Magna* were a mathematical triumph, the culmination of a story that spanned millennia. The Babylonians knew how to solve quadratic equations around 1500 BC, perhaps earlier. The ancient Greeks and Omar Khayyam knew geometric methods for solving cubics, but algebraic solutions of cubic equations, let alone quartics, were unprecedented. At a stroke, mathematics outstripped its classical origins.

There was one tiny snag, however. Cardano noticed it, and several people tried to explain it; they all failed. Sometimes the method works brilliantly; at other times, the formula is as enigmatic as the Delphic oracle. Suppose we apply Cardano's formula to the equation $x^3 - 15x - 4 = 0$. The result is

$$x = \sqrt[3]{2 + \sqrt{-121}} + \sqrt[3]{2 - \sqrt{-121}}$$

However, -121 is negative, so it has no square root. To compound the mystery, there is a perfectly good solution, $x = 4$. The formula doesn't give it.

Light of a kind was shed in 1572 when Bombelli published *L'Algebra*. His main aim was to clarify Cardano's book, but when he came to this particular thorny issue he spotted something Cardano had missed. If you ignore what the symbols mean, and just perform routine calculations, the standard rules of algebra show that

$$(2 + \sqrt{-1})^3 = 2 + \sqrt{-121}$$

Therefore you are entitled to write

$$\sqrt[3]{2 + \sqrt{-121}} = 2 + \sqrt{-1}$$

Similarly,

$$\sqrt[3]{2 - \sqrt{-121}} = 2 - \sqrt{-1}$$

Now the formula that baffled Cardano can be rewritten as

$$(2 + \sqrt{-1}) + (2 - \sqrt{-1})$$

which is equal to 4 because the troublesome square roots cancel out. So Bombelli's nonsensical formal calculations *got the right answer*. And that was a perfectly normal real number.

Somehow, pretending that square roots of negative numbers made sense, even though they obviously did not, could lead to sensible answers. Why?

To answer this question, mathematicians had to develop good ways to think about square roots of negative quantities, and do calculations with them. Early writers, among them Descartes and Newton, interpreted these 'imaginary' numbers as a sign that a problem has no solutions. If you wanted to find a number whose square was minus one, the formal solution 'square root of minus one' was imaginary, so no solution existed. But Bombelli's calculation implied that there was more to imaginaries than that. They could be used to *find* solutions; they could arise as part of the calculation of solutions that *did* exist.

Leibniz had no doubt about the importance of imaginary numbers. In 1702 he wrote: 'The Divine Spirit found a sublime outlet in that wonder of analysis, that portent of the ideal world, that amphibian between being and non-being, which we call the imaginary root of negative unity.' But the eloquence of his statement fails to obscure a fundamental problem: he didn't have a clue what imaginary numbers actually were.

One of the first people to come up with a sensible representation of complex numbers was Wallis. The image of real numbers lying along a line, like marked points on a ruler, was already commonplace. In 1673 Wallis suggested that a complex number $x + iy$ should be thought of as a point in a plane. Draw a line in the plane, and identify points on this line with real numbers in the usual way. Then think of $x + iy$ as a point lying to one side of the line, distance y away from the point x.

Wallis's idea was largely ignored, or worse, criticised. François Daviet de Foncenex, writing about imaginaries in 1758, said that thinking of imaginaries as forming a line at right angles to the real line was pointless. But eventually the idea was revived in a slightly more explicit form. In fact, three people came up with exactly the same method for representing complex numbers, at intervals of a few years, Figure 18. One was a Norwegian surveyor, one a French mathematician, and one a German mathematician. Respectively, they were Caspar Wessel, who published in 1797, Jean-Robert Argand in 1806, and Gauss in 1811. They basically said the same as Wallis, but they added a second line to the picture, an imaginary axis at right angles to the real one. Along this second axis lived the imaginary numbers i, 2i, 3i, and so on. A general complex number, such as 3 + 2i, lived out in the plane, three units along the real axis and two along the imaginary one.

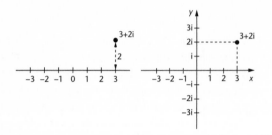

Fig 18 The complex plane. *Left*: according to Wallis. *Right*: according to Wessel, Argand, and Gauss.

This geometric representation was all very well, but it didn't explain why complex numbers form a logically consistent system. It didn't tell us in what sense they are *numbers*. It just provided a way to visualise them. This no more defined what a complex number *is* than a drawing of a straight line defines a real number. It did provide some sort of psychological prop, a slightly artificial link between those crazy imaginaries and the real world, but nothing more.

What convinced mathematicians that they should take imaginary numbers seriously wasn't a logical description of what they were. It was overwhelming evidence that whatever they were, mathematics could make good use of them. You don't ask difficult questions about the philosophical basis of an idea when you are using it every day to solve problems and you can see that it gives the right answers. Foundational questions still have some interest, of course, but they take a back seat to the pragmatic issues of using the new idea to solve old and new problems.

Imaginary numbers, and the system of complex numbers that they spawned, cemented their place in mathematics when a few pioneers turned their attention to complex analysis: calculus (Chapter 3) but with complex numbers instead of real ones. The first step was to extend all the usual functions – powers, logarithms, exponentials, trigonometric functions – to the complex realm. What is $\sin z$ when $z = x + iy$ is complex? What is e^z or $\log z$?

Logically, these things can be whatever we wish. We are operating in a new domain where the old ideas don't apply. It doesn't make much sense, for instance, to think of a right-angled triangle whose sides have complex lengths, so the geometric definition of the sine function is irrelevant. We could take a deep breath, insist that $\sin z$ has its usual value when z is real, but equals 42 whenever z isn't real: job done. But that would be a pretty silly definition: not because it's imprecise, but because it bears no sensible relationship to the original one for real numbers. One requirement for an extended definition must be that it agrees with the old one when applied to real numbers, but that's not enough. It's true for my silly extension of the sine. Another requirement is that the new concept should retain as many features of the old one as we can manage; it should somehow be 'natural'.

What properties of sine and cosine do we want to preserve? Presumably we'd like all the pretty formulas of trigonometry to remain valid, such as $\sin 2z = 2 \sin z \cos z$. This imposes a constraint but doesn't help. A more interesting property, derived using analysis (the rigorous formulation of calculus), is the existence of an infinite series:

$$\sin z = z - \frac{z^3}{1.2.3} + \frac{z^5}{1.2.3.4.5} - \frac{z^7}{1.2.3.4.5.6.7} + \cdots$$

(The sum of such a series is defined to be the limit of the sum of finitely many terms as the number of terms increases indefinitely.) There is a similar series for the cosine:

$$\cos z = 1 - \frac{z^2}{1.2} + \frac{z^4}{1.2.3.4} - \frac{z^6}{1.2.3.4.5.6} + \cdots$$

and the two are obviously related in some way to the series for the exponential:

$$e^z = 1 + z + \frac{z^2}{1.2} + \frac{z^3}{1.2.3} + \frac{z^4}{1.2.3.4} + \cdots$$

These series may seem complicated, but they have an attractive feature: we know how to make sense of them for complex numbers. All they involve is integer powers (which we obtain by repeated multiplication) and a technical issue of convergence (making sense of the infinite sum). Both of these extend naturally into the complex realm and have all of the expected properties. So we can define sines and cosines of complex numbers using the same series that work in the real case.

Since all of the usual formulas in trigonometry are consequences of these series, those formulas automatically carry over as well. So do the basic facts of calculus, such as 'the derivative of sine is cosine'. So does $e^{z+w} = e^z e^w$. This is all so pleasant that mathematicians were happy to settle on the series definitions. And once they'd done that, a great deal else necessarily had to fit in with it. If you followed your nose, you could discover where it led.

For example, those three series look very similar. Indeed, if you replace z by iz in the series for the exponential, you can split the resulting series into two parts, and what you get are precisely the series for sine and cosine. So the series definitions imply that

$$e^{iz} = \cos z + i \sin z.$$

You can also express both sine and cosine using exponentials:

$$\cos z = \frac{e^{iz} + e^{-iz}}{2} \qquad \sin z = \frac{e^{iz} - e^{-iz}}{2i}$$

This hidden relationship is extraordinarily beautiful. But you'd never suspect anything like it could exist if you remained stuck in the realm of the reals. Curious similarities between trigonometric formulas and exponential ones (for example, their infinite series) would remain just that. Viewed through complex spectacles, everything suddenly slots into place.

One of the most beautiful, yet enigmatic, equations in the whole of mathematics emerges almost by accident. In the trigonometric series, the number z (when real) has to be measured in radians, for which a full circle of 360° becomes 2π radians. In particular, the angle 180° is π radians. Moreover, $\sin \pi = 0$ and $\cos \pi = -1$. Therefore

$$e^{i\pi} = \cos \pi + i \sin \pi = -1$$

The imaginary number i unites the two most remarkable numbers in mathematics, e and π, in a single elegant equation. If you've never seen this before, and have any mathematical sensitivity, the hairs on your neck raise and prickles run down your spine. This equation, attributed to Euler, regularly comes top of the list in polls for the most beautiful equation in mathematics. That doesn't mean that it *is* the most beautiful equation, but it does show how much mathematicians appreciate it.

Armed with complex functions and knowing their properties, the mathematicians of the nineteenth century discovered something remarkable: they could use these things to solve differential equations in mathematical physics. They could apply the method to static electricity, magnetism, and fluid flow. Not only that: it was *easy*.

In Chapter 3 we talked of functions – mathematical rules that assign, to any given number, a corresponding number, such as its square or sine. Complex functions are defined in the same way, but now we allow the numbers involved to be complex. The method for solving differential equations was delightfully simple. All you had to do was take some complex function, call it $f(z)$, and split it into its real and imaginary parts:

$$f(z) = u(z) + iv(z)$$

Now you have two *real*-valued functions u and v, defined for any z in the complex plane. Moreover, whatever function you start with, these two component functions satisfy differential equations found in physics. In a fluid-flow interpretation, for example, u and v determine the flow-lines. In an electrostatic interpretation, the two components determine the electric field and how a small charged particle would move; in a magnetic interpretation, they determine the magnetic field and the lines of force.

I'll give just one example: a bar magnet. Most of us remember seeing a famous experiment in which a magnet is placed beneath a sheet of paper, and iron filings are scattered over the paper. They automatically line up to show the lines of magnetic force associated with the magnet – the paths that a tiny test magnet would follow if placed in the magnetic field. The curves look like Figure 19 (*left*).

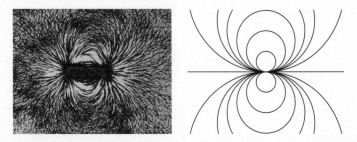

Fig 19 *Left*: Magnetic field of bar magnet. *Right*: Field derived using complex analysis.

To obtain this picture using complex functions, we just let $f(z) = 1/z$. The lines of force turn out to be circles, tangent to the real axis, as in Figure

19 (*right*). This is what the magnetic fields lines of a very tiny bar magnet would look like. A more complicated choice of function corresponds to a magnet of finite size: I chose this function to keep everything as simple as possible.

This was wonderful. There were endless functions to work with. You decided which function to look at, found its real and imaginary parts, worked out their geometry ... and, lo and behold, you had solved a problem in magnetism, or electricity, or fluid flow. Experience soon told you which function to use for which problem. The logarithm was a point source, minus the logarithm was a sink through which fluid disappeared like the plughole in a kitchen sink, i times the logarithm was a point vortex where the fluid spun round and round... It was magic! Here was a method that could churn out solution after solution to problems that would otherwise be opaque. Yet it came with a guarantee of success, and if you were worried about all that complex analysis stuff, you could check directly that the results you obtained really did represent solutions.

This was just the beginning. As well as special solutions, you could prove general principles, hidden patterns in the physical laws. You could analyse waves and solve differential equations. You could transform shapes into other shapes, using complex equations, and the same equations transformed the flow-lines round them. The method was limited to systems in the plane, because that was where a complex number naturally lived, but the method was a godsend when previously even problems in the plane were out of reach. Today, every engineer is taught how to use complex analysis to solve practical problems, early in their university course. The Joukowski transformation $z + 1/z$ turns a circle into an aerofoil shape, the cross-section of a rudimentary aeroplane wing, see Figure 20. It therefore turns the flow past a circle, easy to find if you knew the tricks of the trade, into the flow past an aerofoil. This calculation, and more realistic improvements, were important in the early days of aerodynamics and aircraft design.

This wealth of practical experience made the foundational issues moot. Why look a gift horse in the mouth? There had to be a sensible meaning for complex numbers – they wouldn't work otherwise. Most scientists and mathematicians were much more interested in digging out the gold than they were in establishing exactly where it had come from and what distinguished it from fools' gold. But a few persisted. Eventually, the Irish mathematician William Rowan Hamilton knocked the whole thing on the

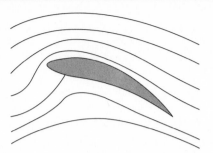

Fig 20 Flow past a wing derived from the Joukowski transformation.

head. He took the geometric representation proposed by Wessel, Argand, and Gauss, and expressed it in coordinates. A complex number was a pair of real numbers (x, y). The real numbers were those of the form $(x, 0)$. The imaginary i was $(0, 1)$. There were simple formulas for adding and multiplying these pairs. If you were worried about some law of algebra, such as the commutative law $ab = ba$, you could routinely work out both sides as pairs, and make sure they were the same. (They were.) If you identified $(x, 0)$ with plain x, you embedded the real numbers into the complex ones. Better still, $x + iy$ then worked out as the pair (x, y).

This wasn't just a representation, but a *definition*. A complex number, said Hamilton, is nothing more nor less than a pair of ordinary real numbers. What made them so useful was an inspired choice of the rules for adding and multiplying them. What they actually were was trite; it was how you used them that produced the magic. With this simple stroke of genius, Hamilton cut through centuries of heated argument and philosophical debate. But by then, mathematicians had become so used to working with complex numbers and functions that no one cared any more. All you needed to remember was that $i^2 = -1$.

6 Much ado about knotting
Euler's Formula for Polyhedra

number of faces

number of edges

number of vertices

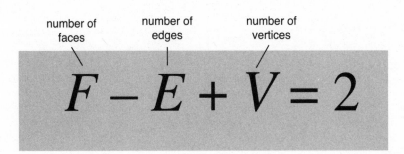

$$F - E + V = 2$$

What does it say?

The numbers of faces, edges, and vertices of a solid are not independent, but are related in a simple manner.

Why is that important?

It distinguishes between solids with different topologies using the earliest example of a topological invariant. This paved the way to more general and more powerful techniques, creating a new branch of mathematics.

What did it lead to?

One of the most important and powerful areas of pure mathematics: topology, which studies geometric properties that are unchanged by continuous deformations. Examples include surfaces, knots, and links. Most applications are indirect, but its influence behind the scenes is vital. It helps us understand how enzymes act on DNA in a cell, and why the motion of celestial bodies can be chaotic.

As the nineteenth century approached its end, mathematicians began to develop a new kind of geometry, one in which familiar concepts such as lengths and angles played no role whatsoever and no distinction was made between triangles, squares, and circles. Initially it was called *analysis situs*, the analysis of position, but mathematicians quickly settled on another name: topology.

Topology has its roots in a curious numerical pattern that Descartes noticed in 1639 when thinking about Euclid's five regular solids. Descartes was a French-born polymath who spent most of his life in the Dutch Republic, present-day Netherlands. His fame mainly rests on his philosophy, which proved so influential that for a long time Western philosophy consisted largely of responses to Descartes. Not always in agreement, you appreciate, but motivated by his arguments nonetheless. His sound bite *cogito ergo sum* – 'I think, therefore I am' – has become common cultural currency. But Descartes's interests extended beyond philosophy into science and mathematics.

In 1639 Descartes turned his attention to the regular solids, and this was when he noticed his curious numerical pattern. A cube has 6 faces, 12 edges, and 8 vertices; the sum $6 - 12 + 8$ equals 2. A dodecahedron has 12 faces, 30 edges, and 20 vertices; the sum $12 - 30 + 20 = 2$. An icosahedron has 20 faces, 30 edges, and 12 vertices; the sum $20 - 30 + 12 = 2$. The same relationship holds for the tetrahedron and octahedron. In fact, it applies to a solid of *any* shape, regular or not. If the solid has F faces, E edges, and V vertices, then $F - E + V = 2$. Descartes viewed this formula as a minor curiosity and did not publish it. Only much later did mathematicians see this simple little equation as one of the first tentative steps towards the great success story in twentieth-century mathematics, the inexorable rise of topology. In the nineteenth century, the three pillars of pure mathematics were algebra, analysis, and geometry. By the end of the twentieth, they were algebra, analysis, and topology.

Topology is often characterised as 'rubber-sheet geometry' because it is the kind of geometry that would be appropriate for figures drawn on a sheet of elastic, so that lines can bend, shrink, or stretch, and circles can be squashed so that they turn into triangles or squares. All that matters is *continuity*: you are not allowed to rip the sheet apart. It may seem remarkable that anything so weird could have any importance, but continuity is a basic aspect of the natural world and a fundamental feature of mathematics. Today we mostly use topology indirectly, as one mathematical technique among many. You don't find anything obviously topological in your kitchen. However, a Japanese company did market a chaotic dishwasher, which according to their marketing people cleaned dishes more efficiently, and our understanding of chaos rests on topology. So do some important aspects of quantum field theory and that iconic molecule DNA. But, when Descartes counted the most obvious features of the regular solids and noticed that they were not independent, all this was far in the future.

It was left to the indefatigable Euler, the most prolific mathematician in history, to prove and publish this relationship, which he did in 1750 and 1751. I'll sketch a modern version. The expression $F - E + V$ may seem fairly arbitrary, but it has a very interesting structure. Faces (F) are polygons, of dimension 2, edges (E) are lines, so have dimension 1, and vertices (V) are points, of dimension 0. The signs in the expression alternate, $+ - +$, with $+$ being assigned to features of even dimension and $-$ to those of odd dimension. This implies that you can simplify a solid by merging its faces or removing edges and vertices, and these changes will not alter the number $F - E + V$ provided that every time you get rid of a face you also remove an edge, or every time you get rid of a vertex you also remove an edge. The alternating signs mean that changes of this kind cancel out.

Now I'll explain how this clever structure makes the proof work. Figure 21 shows the key stages. Take your solid. Deform it into a nice round sphere, with its edges being curves on that sphere. If two faces meet along a common edge, then you can remove that edge and merge the faces into one. Since this merger reduces both F and E by 1, it doesn't change $F - E + V$. Keep doing this until you get down to a single face, which covers almost all of the sphere. Aside from this face, you are left with only edges and vertices. These must form a tree, a network with no closed loops, because any closed loop on a sphere separates at least two faces: one inside it, the other outside it. The branches of this tree are the remaining edges of the solid, and they join together at the remaining vertices. At this stage only one face remains: the entire sphere, minus the tree. Some branches of this tree connect to other branches at both ends, but some, at the extremes, terminate in a vertex, to which no other branches attach. If you remove one of these terminating branches together with that vertex, then the tree gets smaller, but since both E and V decrease by 1, $F - E + V$ again remains unchanged.

This process continues until you are left with a single vertex sitting on an otherwise featureless sphere. Now $V = 1$, $E = 0$, and $F = 1$. So $F - E + V = 1 - 0 + 1 = 2$. But since each step leaves $F - E + V$ unchanged, its value at the beginning must also have been 2, which is what we want to prove.

Fig 21 Key stages in simplifying a solid. *Left to right*: (1) Start. (2) Merging adjacent faces. (3) Tree that remains when all faces have been merged. (4) Removing an edge and a vertex from the tree. (5) End.

It's a cunning idea, and it contains the germ of a far-reaching principle. The proof has two ingredients. One is a simplification process: remove either a face and an adjacent edge or a vertex and an edge that meets it. The other is an invariant, a mathematical expression that remains unchanged whenever you carry out a step in the simplification process. Whenever these two ingredients coexist, you can compute the value of the invariant for any initial object by simplifying it as far as you can, and then computing the value of the invariant for this simplified version. Because it is an invariant, the two values must be equal. Because the end result is simple, the invariant is easy to calculate.

Now I have to admit that I've been keeping one technical issue up my sleeve. Descartes's formula does not, in fact, apply to any solid. The most familiar solid for which it fails is a picture frame. Think of a picture frame made from four lengths of wood, each rectangular in cross-section, joined at the four corners by 45° mitres as in Figure 22 (*left*). Each length of wood contributes 4 faces, so $F = 16$. Each length also contributes 4 edges, but the mitre joint creates 4 more at each corner, so $E = 32$. Each corner comprises 4 vertices, so $V = 16$. Therefore $F - E + V = 0$.

What went wrong?

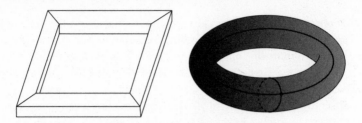

Fig 22 *Left*: A picture frame with $F-E+V=0$. *Right*: Final configuration when the picture frame is smoothed and then simplified.

There's no problem with $F-E+V$ being invariant. Neither is there much of a problem with the simplification process. But if you work through it for the frame, always cancelling one face against one edge, or one vertex against one edge, then the final simplified configuration is not a single vertex sitting in a single face. Performing the cancellation in the most obvious way, what you get is Figure 22 (*right*), with $F=1$, $V=1$, $E=2$. I've smoothed the faces and edges for reasons that will quickly become apparent. At this stage removing an edge just merges the sole remaining face with itself, so the changes to the numbers no longer cancel. This is why we stop, but we're home and dry anyway: for this configuration, $F-E+V=0$. So the *method* performs perfectly. It just yields a different result for the picture frame. There must be some fundamental difference between a picture frame and a cube, and the invariant $F-E+V$ is picking it up.

The difference turns out to be a topological one. Early in my version of Euler's proof, I told you to take the solid and 'deform it into a nice round sphere'. But this is not possible for the picture frame. It's not shaped like a sphere, even after being simplified. It is a torus, which looks like an inflatable rubber ring with a hole through the middle. The hole is also clearly visible in the original shape: it's where the picture would go. A sphere, in contrast, has no holes. The hole in the frame is why the simplification process leads to a different result. However, we can wrest victory from the jaws of defeat, because $F-E+V$ is still an invariant. So the proof tells us that *any* solid that is deformable into a torus will satisfy the slightly different equation $F-E+V=0$. In consequence, we have the basis of a rigorous proof that a torus cannot be deformed into a sphere: that is, the two surfaces are topologically different.

Of course this is intuitively obvious, but now we can support intuition with logic. Just as Euclid started from obvious properties of points and lines, and formalised them into a rigorous theory of geometry, the mathematicians of the nineteenth and twentieth centuries could now develop a rigorous formal theory of topology.

Fig 23 *Left*: 2-holed torus. *Right*: 3-holed torus.

Where to start was a no-brainer. There exist solids like a torus but with two or more holes, as in Figure 23, and the same invariant should tell us something useful about those. It turns out that any solid deformable into a 2-holed torus satisfies $F - E + V = -2$, any solid deformable into a 3-holed torus satisfies $F - E + V = -4$, and in general any solid deformable into a g-holed torus satisfies $F - E + V = 2 - 2g$. The symbol g is short for 'genus', the technical name for 'number of holes'. Pursuing the line of thought that Descartes and Euler began leads to a connection between a quantitative property of solids, the number of faces, vertices, and edges, and a qualitative property, possessing holes. We call $F - E + V$ the Euler characteristic of the solid, and observe that it depends only on which solid we are considering and not on how we cut it into faces, edges, and vertices. This makes it an intrinsic feature of the solid itself.

Agreed, we count the number of holes, a quantitative operation, but 'hole' itself is qualitative in the sense that it's not obviously a feature of the solid at all. Intuitively, it's a region in space where the solid *isn't*. But not any such region. After all, that description applies to all of the space surrounding the solid, and no one would consider it all to be a hole. And it also applies to all of the space surrounding a sphere ... which doesn't *have* a hole. In fact, the more you start to think about what a hole is, the more you realise that it's quite tricky to define one. My favourite example to show just how confusing it all gets is the shape in Figure 24, known as a hole-through-a-hole-in-a-hole. Apparently you can thread a hole through another hole, which is actually a hole in a third hole.

This way lies madness.

It wouldn't much matter if solids with holes in them never turned up anywhere important. But by the end of the nineteenth century they were turning up all over mathematics – in complex analysis, algebraic geometry, and Riemann's differential geometry. Worse, higher-dimensional analogues of solids were taking centre stage, in all areas of pure and

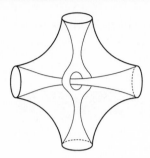

Fig 24 Hole-through-a-hole-in-a-hole.

applied mathematics; as already noted, the dynamics of the Solar System requires 6 dimensions per body. And they had higher-dimensional analogues of holes. Somehow it was necessary to bring a modicum of order into the area. And the answer turned out to be ... invariants.

The idea of a topological invariant goes back to Gauss's work on magnetism. He was interested in how magnetic and electrical field lines could link with each other, and he defined the linking number, which counts how many times one field line winds round another. This is a topological invariant: it remains the same if the curves are continuously deformed. He found a formula for this number using integral calculus, and every so often he expressed a wish for a better understanding of the 'basic geometric properties' of diagrams. It is no coincidence that the first serious inroads into such an understanding came through the work of one of Gauss's students, Johann Listing, and Gauss's assistant August Möbius. Listing's *Vorstudien zur Topologie* ('Studies in Topology') of 1847 introduced the word 'topology', and Möbius made the role of continuous transformations explicit.

Listing had a bright idea: seek generalisations of Euler's formula. The expression $F - E + V$ is a combinatorial invariant: a feature of a specific way of describing a solid, based on cutting it into faces, edges, and vertices. The number g of holes is a topological invariant: something that does not change however the solid is deformed, as long as the deformation is continuous. A topological invariant captures a qualitative conceptual feature of a shape; a combinatorial one provides a method for calculating it. The two together are very powerful, because we can use the conceptual invariant to think about shapes, and the combinatorial version to pin down what we are talking about.

In fact, the formula lets us sidestep the tricky issue of defining 'hole' altogether. Instead, we define 'number of holes' as a package, without either defining a hole or counting how many there are. How? Easy. Just rewrite the generalised version of Euler's formula $F-E+V=2-2g$ in the form

$$g = 1 - F/2 + E/2 - V/2$$

Now we can calculate g by drawing faces and so forth on our solid, counting F, E, and V, and substituting those values into the formula. Since the expression is an invariant, it doesn't matter how we cut the solid up: we always get the same answer. But nothing that we do depends on having a definition of a hole. Instead, 'number of holes' becomes an interpretation, in intuitive terms, derived by looking at simple examples where we feel we know what the phrase should mean.

It may seem like a cheat, but it makes significant inroads into a central question in topology: when can one shape be continuously deformed into another? That is, as far as topologists are concerned, are the two shapes the same or not? If they are the same, their invariants must also be the same; conversely, if the invariants are different, so are the shapes. (However, sometimes two shapes might have the same invariant, but be different; it depends on the invariant.) Since a sphere has Euler characteristic 2, but a torus has Euler characteristic 0, there is no way to deform a sphere continuously into a torus. This may seem obvious, because of the hole... but we've seen the turbulent waters into which that way of thinking can lead. You don't have to interpret the Euler characteristic in order to use it to distinguish shapes, and here it is decisive.

Less obviously, the Euler characteristic shows that the puzzling hole-through-a-hole-in-a-hole (Figure 24) is actually just a 3-holed torus in disguise. Most of the apparent complexity stems not from the intrinsic topology of the surface, but from the way I have chosen to embed it in space.

The first really significant theorem in topology grew out of the formula for the Euler characteristic. It was a complete classification of surfaces, curved two-dimensional shapes like the surface of a sphere or that of a torus. A couple of technical conditions were also imposed: the surface should have no boundary, and it should be of finite extent (the jargon is 'compact').

For this purpose a surface is described intrinsically; that is, it is not conceived as existing in some surrounding space. One way to do this is to view the surface as a number of polygonal regions (which topologically are equivalent to circular discs) that are glued together along their edges according to specified rules, like the 'glue tab A to tab B' instructions you get when assembling a cardboard cut-out. A sphere, for instance, can be described using two discs, glued together along their boundaries. One disc becomes the northern hemisphere, the other the southern hemisphere. A torus has an especially elegant description as a square with opposite edges glued to each other. This construction can be visualised in a surrounding space (Figure 25), which explains why it creates a torus, but the mathematics can be carried out using just the square together with the gluing rules, and this offers advantages precisely because it is intrinsic.

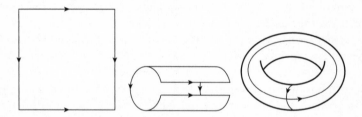

Fig 25 Gluing the edges of a square to make a torus.

The possibility of gluing bits of boundary together leads to a rather strange phenomenon: surfaces with only one side. The most famous example is the Möbius band, introduced by Möbius and Listing in 1858, which is a rectangular strip whose ends are glued together with a 180° turn (usually called a half-twist, on the convention that 360° constitutes a full twist). The Möbius band, see Figure 26 (*left*), has an edge, comprising the edges of the rectangle that don't get glued to anything. This is the only edge, because the two separate edges of the rectangle are connected together into a closed loop by the half-twist, which glues them end to end.

It is possible to make a model of a Möbius band from paper, because it embeds naturally in three-dimensional space. The band has only one side, in the sense that if you start painting one of its surfaces, and keep going, you will eventually cover the entire surface, front and back. This happens because the half-twist connects the front to the back. That's not an intrinsic description, because it relies on embedding the band in space, but there is an equivalent, more technical property known as orientability, which is intrinsic.

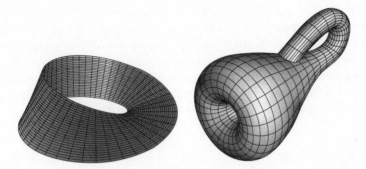

Fig 26 *Left*: Möbius band. *Right*: Klein bottle. The apparent self-intersection occurs because the drawing embeds it in three-dimensional space.

There is a related surface with only one side, having no edges at all, Figure 26 (*right*). It arises if we glue two sides of a rectangle together like a Möbius band, and glue the other two sides together without any twisting. Any model in three-dimensional space has to pass through itself, even though from an intrinsic point of view the gluing rules do not introduce any self-intersections. If this surface is pictured with such a crossing, it looks like a bottle whose neck has been poked through the side wall and joined to the bottom. It was invented by Felix Klein, and is known as a Klein bottle – almost certainly a joke based on a German pun, changing *Kleinsche Fläche* (Klein's surface) to *Kleinsche Flasche* (Klein's bottle).

The Klein bottle has no boundary and is compact, so any classification of surfaces must include it. It is the best known of an entire family of one-sided surfaces, and surprisingly it is not the simplest. This honour goes to the projective plane, which arises if you glue both pairs of opposite sides of a square together, with a half-twist for each. (This is difficult to do with paper because paper is too rigid; like the Klein bottle it requires the surface to intersect itself. It is best done 'conceptually', that is, by drawing pictures on the square but remembering the gluing rules when lines go off the edge and 'wrap round'.) The classification theorem for surfaces, proved by Johann Listing around 1860, leads to two families of surfaces. Those with two sides are the sphere, torus, 2-holed torus, 3-holed torus, and so on. Those with only one side form a similar infinite family, starting with the projective plane and the Klein bottle. They can be obtained by cutting a small disc out of the corresponding two-sided surface and gluing in a Möbius band instead.

Surfaces turn up naturally in many areas of mathematics. They are important in complex analysis, where surfaces are associated with singularities, points at which functions behave strangely – for instance, the derivative fails to exist. Singularities are the key to many problems in complex analysis; in a sense they capture the essence of the function. Since singularities are associated with surfaces, the topology of surfaces provides an important technique for complex analysis. Historically, this motivated the classification.

Most modern topology is highly abstract, and a lot of it happens in four or more dimensions. We can get a feel for the subject in a more familiar setting: knots. In the real world, a knot is a tangle tied in a length of string. Topologists need a way to stop the knot escaping off the ends once it has been tied, so they join the ends of the string together to form a closed loop. Now a knot is just a circle embedded in space. Intrinsically, a knot is topologically identical to a circle, but on this occasion what counts is how the circle sits inside its surrounding space. This might seem contrary to the spirit of topology, but the essence of a knot lies in the relation between the loop of string and the space that surrounds it. By considering not just the loop, but how it relates to space, topology can tackle important questions about knots. Among these are:

- How do we know a knot is really knotted?
- How can we distinguish topologically different knots?
- Can we classify all possible knots?

Experience tells us that there are many different types of knot. Figure 27 shows a few of them: the overhand or trefoil knot, reef knot, granny knot, figure-8, stevedore's knot, and so on. There is also the unknot, an ordinary circular loop; as the name reflects, this loop is *not* knotted. Many different kinds of knot have been used by generations of mariners, mountaineers, and boy scouts. Any topological theory should of course reflect this wealth of experience, but everything has to be proved, rigorously, within the formal setting of topology, just as Euclid had to prove Pythagoras's theorem instead of just drawing a few triangles and measuring them. Remarkably, the first topological proof that knots exist, in the sense that there is an embedding of the circle that cannot be deformed into the

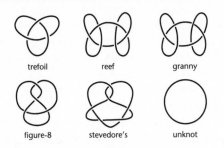

| trefoil | reef | granny |
| figure-8 | stevedore's | unknot |

Fig 27 Five knots and the unknot.

unknot, first appeared in 1926 in the German mathematician Kurt Reidemeister's *Knoten und Gruppen* ('Knots and Groups'). The word 'group' is a technical term in abstract algebra, which quickly became the most effective source of topological invariants. In 1927 Reidemeister, and independently the American James Waddell Alexander, in collaboration with his student G. B. Briggs, found a simpler proof of the existence of knots using the 'knot diagram'. This is a cartoon image of the knot, drawn with tiny breaks in the loop to show how the separate strands overlap, as in Figure 27. The breaks are not present in the knotted loop itself, but they represent its three-dimensional structure in a two-dimensional diagram. Now we can use the breaks to split the knot diagram into a number of distinct pieces, its components, and then we can manipulate the diagram and see what happens to the components.

If you look back at how I used the invariance of the Euler characteristic, you'll see that I simplified the solid using a series of special moves: merge two faces by removing an edge, merge two edges by removing a point. The same trick applies to knot diagrams, but now you need three types of move to simplify them, called Reidemeister moves, Figure 28. Each move can be carried out in either direction: add or remove a twist, overlap two strands or pull them apart, move one strand through the place where two others cross.

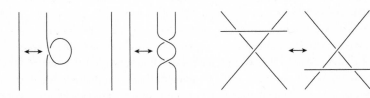

Fig 28 Reidemeister moves.

With some preliminary fiddling to tidy up the knot diagram, such as modifying places where three curves overlap if that ever happens, it can be proved that any deformation of a knot can be represented as a finite series of Reidemeister moves applied to its diagram. Now we can play the Euler game; all we have to do is find an invariant. Among them is the knot group, but there is a far simpler invariant that proves the trefoil really is a knot. I can explain it in terms of colouring the separate components in a knot diagram. I'm starting with a slightly more complicated diagram than I have to, with an extra loop, in order to illustrate some features of the idea, Figure 29.

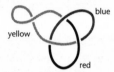

Fig 29 Colouring a trefoil knot with an extra twist.

The extra twist creates four separate components. Suppose I colour the components using three colours for each, say red, yellow, and blue (shown in the figure as black, light grey, and dark grey). Then this colouring obeys two simple rules:

- At least two distinct colours are used. (Actually all three are, but that's extra information that I don't need.)
- At each crossing, either the three strands near the crossing all have different colours or they are all the same colour. Near the crossing caused by my extra loop, all three components are yellow. Two of these components (in yellow) join up elsewhere, but near the crossing they are separate.

The wonderful observation is that if a knot diagram can be coloured using three colours, obeying these two rules, then the same is true after any Reidemeister move. You can prove this very easily by working out how the Reidemeister moves affect the colours. For example, if I untwist the extra loop in my picture then I can leave the colours unchanged and everything still works. Why is this wonderful? Because it proves that the trefoil really is knotted. Suppose for the sake of argument that it can be unknotted; then some series of Reidemeister moves turns it into an unknotted loop. Since the trefoil can be coloured to obey the two rules, the same must apply to the unknotted loop. But an unknotted loop consists of a single strand with no overlaps, so the only way to colour it is to use the same colour everywhere. But this violates the first rule. By contradiction, no such series of Reidemeister moves can exist; that is, the trefoil can't be unknotted.

This proves that the trefoil is knotted, but doesn't distinguish it from other knots such as the reef knot or the stevedore's knot. One of the earliest effective ways to do this was invented by Alexander. It was derived from Reidemeister's abstract algebra methods, but it leads to an invariant that is algebraic in the more familiar sense of school algebra. It's called the Alexander polynomial, and it associates to any knot a formula formed from powers of a variable x. Strictly speaking, the term 'polynomial' applies only when the powers are positive integers, but here we also allow negative powers. Table 2 lists a few Alexander polynomials. If two knots in the list have different Alexander polynomials, and here all but the reef and granny do, then the knots must be topologically different. The converse is not true: the reef and granny have the same Alexander polynomials, but in 1952 Ralph Fox proved that they are topologically different. The proof required surprisingly sophisticated topology. It was far more difficult than anyone expected.

knot	Alexander polynomial
Unknot	1
Trefoil	$x - 1 + x^{-1}$
Figure-8	$-x + 3 - x^{-1}$
Reef	$x^2 - 2x + 3 - 2x^{-1} + x^{-2}$
Granny	$x^2 - 2x + 3 - 2x^{-1} + x^{-2}$
Stevedore's knot	$-2x + 5 - 2x^{-1}$

Table 2 Alexander polynomials of knots.

After about 1960 knot theory entered the topological doldrums, becalmed in a vast ocean of unsolved questions, awaiting a breath of creative insight. It came in 1984, when the New Zealand mathematician Vaughan Jones had an idea so simple that it could have occurred to anyone from Reidemeister onwards. Jones wasn't a knot theorist; he wasn't even a topologist. He was an analyst, working on operator algebras, an area with strong links to mathematical physics. It wasn't a total surprise that the ideas applied to knots, because mathematicians and physicists already knew of interesting connections between operator algebras and braids, which are a special kind of multi-stranded knot. The new knot invariant he invented, called the Jones polynomial, is also defined using the knot diagram and three types of move. However, the moves do not preserve the topological type of the knot; they do not preserve the new 'Jones polynomial'. Amazingly, however, the idea can still be made to work, and the Jones polynomial is a knot invariant.

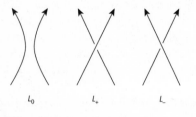

Fig 30 Jones moves.

For this invariant we have to choose a specific direction along the knot, shown by an arrow. The Jones polynomial $V(x)$ is defined to be one for the unknot. Given any knot L_0, move two separate strands close together without changing any crossings in its diagram. Be careful to align the directions as shown: that's why the arrow is needed, and the process doesn't work without it. Replace that region of L_0 by two strands that cross in the two possible ways (Figure 30). Let the resulting knot diagrams be L_+ and L_-. Now define

$$(x^{1/2} - x^{-1/2})V(L_0) = x^{-1}V(L_+) - xV(L_-)$$

By starting with the unknot and applying such moves in the right way, you can work out the Jones polynomial for any knot. Mysteriously, it turns out to be a topological invariant. And it outperforms the traditional Alexander polynomial; for instance, it can distinguish reef from granny, because they have different Jones polynomials.

Jones's discovery won him the Fields medal, the most prestigious prize in mathematics. It also triggered an outburst of new knot invariants. In 1985 four different groups of mathematicians, eight people in total, simultaneously discovered the same generalisation of the Jones polynomial and submitted their papers independently to the same journal. All four proofs were different, and the editor persuaded the eight authors to join forces and publish one combined article. Their invariant is often called the HOMFLY polynomial, based on their initials. But even the

Jones and HOMFLY polynomials have not fully answered the three problems of knot theory. It is not known whether a knot with Jones polynomial 1 must be unknotted, though many topologists think this is probably true. There exist topologically distinct knots with the same Jones polynomial; the simplest examples known have ten crossings in their knot diagrams. A systematic classification of all possible knots remains a mathematician's pipedream.

It's pretty, but is it useful? Topology has many uses, but they are usually indirect. Topological principles provide insight into other, more directly applicable, areas. For instance, our understanding of chaos is founded on topological properties of dynamical systems, such as the bizarre behaviour that Poincaré noted when he rewrote his prizewinning memoir (Chapter 4). The Interplanetary Superhighway is a topological feature of the dynamics of the Solar System.

More esoteric applications of topology arise at the frontiers of fundamental physics. Here the main consumers of topology are quantum field theorists, because the theory of superstrings, the hoped-for unification of quantum mechanics and relativity, is based on topology. Here analogues of the Jones polynomial in knot theory arise in the context of Feynman diagrams, which show how quantum particles such as electrons and photons move through space-time, colliding, merging, and breaking apart. A Feynman diagram is a bit like a knot diagram, and Jones's ideas can be extended to this context.

To me one of the most fascinating applications of topology is its growing use in biology, helping us to understand the workings of the molecule of life, DNA. Topology turns up because DNA is a double helix, like two spiral staircases winding around each other. The two strands are intricately intertwined, and important biological processes, in particular the way a cell copies its DNA when it divides, have to take account of this complex topology. When Francis Crick and James Watson published their work on the molecular structure of DNA in 1953 they ended with a brief allusion to a possible copying mechanism, presumably involved in cell division, in which the two strands were pulled apart and each was used as the template for a new copy. They were reluctant to claim too much, because they were aware that there are topological obstacles to pulling apart intertwined strands. Being too specific about their proposal might have muddied the waters at such an early stage.

As things turned out, Crick and Watson were right. The topological obstacles are real, but evolution has provided methods for overcoming them, such as special enzymes that cut-and-paste strands of DNA. It is no coincidence that one of these is called topoisomerase. In the 1990s mathematicians and molecular biologists used topology to analyse the twists and turns of DNA, and to study how it works in the cell, where the usual method of X-ray diffraction can't be used because it requires the DNA to be in crystalline form.

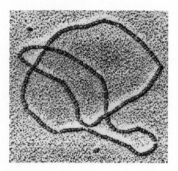

Fig 31 Loop of DNA forming a trefoil knot.

Some enzymes, called recombinases, cut the two DNA strands and rejoin them in a different way. To determine how such an enzyme acts when it is in a cell, biologists apply the enzyme to a closed loop of DNA. Then they observe the shape of the modified loop using an electron microscope. If the enzyme joins distinct strands together, the image is a knot, Figure 31. If the enzyme keeps the strands separate, the image shows two linked loops. Methods from knot theory, such as the Jones polynomial and another theory known as 'tangles', make it possible to work out which knots and links occur, and this provides detailed information about what the enzyme does. They also make new predictions that have been verified experimentally, giving some confidence that the mechanism indicated by the topological calculations is correct.[1]

One the whole, you won't run into topology in everyday life, aside from that dishwasher I mentioned at the start of this chapter. But behind the scenes, topology informs the whole of mainstream mathematics, enabling the development of other techniques with more obvious practical uses. This is why mathematicians consider topology to be of vast importance, while the rest of the world has hardly heard of it.

7

Patterns of chance
Normal Distribution

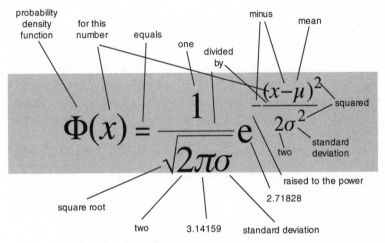

probability density function · for this number · equals · one divided by · minus · mean

$$\Phi(x) = \frac{1}{\sqrt{2\pi\sigma}} e^{-\frac{(x-\mu)^2}{2\sigma^2}}$$

squared · standard deviation · two · raised to the power · 2.71828 · square root · two · 3.14159 · standard deviation

What does it say?

The probability of observing a particular data value is greatest near the mean value – the average – and dies away rapidly as the difference from the mean increases. How rapidly depends on a quantity called the standard deviation.

Why is that important?

It defines a special family of bell-shaped probability distributions, which are often good models of common real-world observations.

What did it lead to?

The concept of the 'average man', tests of the significance of experimental results, such as medical trials, and an unfortunate tendency to default to the bell curve as if nothing else existed.

Mathematics is about patterns. The random workings of chance seem to be about as far from patterns as you can get. In fact, one of the current definitions of 'random' boils down to 'lacking any discernible pattern'. Mathematicians had been investigating patterns in geometry, algebra, and analysis for centuries before they realised that even randomness has its own patterns. But the patterns of chance do not conflict with the idea that random events have no pattern, because the regularities of random events are statistical. They are features of a whole series of events, such as the average behaviour over a long run of trials. They tell us nothing about which event occurs at which instant. For example, if you throw a dice[1] repeatedly, then about one sixth of the time you will roll 1, and the same holds for 2, 3, 4, 5, and 6 – a clear statistical pattern. But this tells you nothing about which number will turn up on the next throw.

Only in the nineteenth century did mathematicians and scientists realise the importance of statistical patterns in chance events. Even human actions, such as suicide and divorce, are subject to quantitative laws, on average and in the long run. It took time to get used to what seemed at first to contradict free will. But today these statistical regularities form the basis of medical trials, social policy, insurance premiums, risk assessments, and professional sports.

And gambling, which is where it all began.

Appropriately, it was all started by the gambling scholar, Girolamo Cardano. Being something of a wastrel, Cardano brought in much-needed cash by taking wagers on games of chess and games of chance. He applied his powerful intellect to both. Chess does not depend on chance: winning depends on a good memory for standard positions and moves, and an intuitive sense of the overall flow of the game. In a game of chance, however, the player is subject to the whims of Lady Luck. Cardano realised that he could apply his mathematical talents to good effect even in this tempestuous relationship. He could improve his performance at games of chance by having a better grasp of the odds – the likelihood of winning or losing – than his opponents did. He put together a book on the topic, *Liber de Ludo Aleae* ('Book on Games of Chance'). It remained unpublished until 1633. Its scholarly content is the first systematic treatment of the mathematics of probability. Its less reputable content is a chapter on how to cheat and get away with it.

One of Cardano's fundamental principles was that in a fair bet, the stakes should be proportional to the number of ways in which each player can win. For example, suppose the players roll a dice, and the first player wins if he throws a 6, while the second player wins if he throws anything else. The game would be highly unfair if each bet the same amount to play the game, because the first player has only one way to win, whereas the second has five. If the first player bets £1 and the second bets £5, however, the odds become equitable. Cardano was aware that this method of calculating fair odds depends on the various ways of winning being equally likely, but in games of dice, cards, or coin-tossing it was clear how to ensure that this condition applied. Tossing a coin has two outcomes, heads or tails, and these are equally likely if the coin is fair. If the coin tends to throw more heads than tails, it is clearly biased – unfair. Similarly the six outcomes for a fair dice are equally likely, as are the 52 outcomes for a card drawn from a pack.

The logic behind the concept of fairness here is slightly circular, because we infer bias from a failure to match the obvious numerical conditions. But those conditions are supported by more than mere counting. They are based on a feeling of symmetry. If the coin is a flat disc of metal, of uniform density, then the two outcomes are related by a symmetry of the coin (flip it over). For dice, the six outcomes are related by symmetries of the cube. And for cards, the relevant symmetry is that no card differs significantly from any other, except for the value written on its face. The frequencies 1/2, 1/6, and 1/52 for any given outcome rest on these basic symmetries. A biased coin or biased dice can be created by the covert insertion of weights; a biased card can be created using subtle marks on the back, which reveal its value to those in the know.

There are other ways to cheat, involving sleight of hand – say, to swap a biased dice into and out of the game before anyone notices that it always throws a 6. But the safest way to 'cheat' – to win by subterfuge – is to be perfectly honest, but to know the odds better than your opponent. In one sense you are taking the moral high ground, but you can improve your chances of finding a suitably naive opponent by rigging not the odds but your opponent's expectation of the odds. There are many examples where the actual odds in a game of chance are significantly different from what many people would naturally assume.

An example is the game of crown and anchor, widely played by British seamen in the eighteenth century. It uses three dice, each bearing not the numbers 1–6 but six symbols: a crown, an anchor, and the four card suits of diamond, spade, club, and heart. These symbols are also marked on a mat. Players bet by placing money on the mat and throwing the three dice. If any of the symbols that they have bet on shows up, the banker pays them their stake, multiplied by the number of dice showing that symbol. For example, if they bet £1 on the crown, and two crowns turn up, they win £2 in addition to their stake; if three crowns turn up, they win £3 in addition to their stake. It all sounds very reasonable, but probability theory tells us that in the long run a player can expect to lose 8% of his stake.

Probability theory began to take off when it attracted the attention of Blaise Pascal. Pascal was the son of a Rouen tax collector and a child prodigy. In 1646 he was converted to Jansenism, a sect of Roman Catholicism that Pope Innocent X deemed heretical in 1655. The year before, Pascal had experienced what he called his 'second conversion', probably triggered by a near-fatal accident when his horses fell off the edge of Neuilly bridge and his carriage nearly did the same. Most of his output from then on was religious philosophy. But just before the accident, he and Fermat were writing to each other about a mathematical problem to do with gambling. The Chevalier de Meré, a French writer who called himself a knight even though he wasn't, was a friend of Pascal's, and he asked how the stakes in a series of games of chance should be divided if the contest had to be abandoned part way through. This question was not new: it goes back to the Middle Ages. What was new was its solution. In an exchange of letters, Pascal and Fermat found the correct answer. Along the way they created a new branch of mathematics: probability theory.

A central concept in their solution was what we now call 'expectation'. In a game of chance, this is a player's average return in the long run. It would, for example, be 92 pence for crown and anchor with a £1 stake. After his second conversion, Pascal put his gambling past behind him, but he enlisted its aid in a famous philosophical argument, Pascal's wager.[2] Pascal assumed, playing Devil's advocate, that someone might consider the existence of God to be highly unlikely. In his *Pensées* ('Thoughts') of 1669, Pascal analysed the consequences from the point of view of probabilities:

Let us weigh the gain and the loss in wagering that God is [exists]. Let us estimate these two chances. If you gain, you gain all; if you lose, you lose nothing. Wager, then, without hesitation that He is... There is here an infinity of an infinitely happy life to gain, a chance of gain against a finite number of chances of loss, and what you stake is finite. And so our proposition is of infinite force, when there is the finite to stake in a game where there are equal risks of gain and of loss, and the infinite to gain.

Probability theory arrived as a fully fledged area of mathematics in 1713 when Jacob Bernoulli published his *Ars Conjectandi* ('Art of Conjecturing'). He started with the usual working definition of the probability of an event: the proportion of occasions on which it will happen, in the long run, nearly all the time. I say 'working definition' because this approach to probabilities runs into trouble if you try to make it fundamental. For example, suppose that I have a fair coin and keep tossing it. Most of the time I get a random-looking sequence of heads and tails, and if I keep tossing for long enough I will get heads roughly half the time. However, I seldom get heads exactly half the time: this is impossible on odd-numbered tosses, for example. If I try to modify the definition by taking inspiration from calculus, so that the probability of throwing heads is the limit of the proportion of heads as the number of tosses tends to infinity, I have to prove that this limit exists. But sometimes it doesn't. For example, suppose that the sequence of heads and tails goes

$$\text{THHTTTHHHHHHTTTTTTTTTTTT...}$$

with one tail, two heads, three tails, six heads, twelve tails, and so on – the numbers doubling at each stage after the three tails. After three tosses the proportion of heads is 2/3, after six tosses it is 1/3, after twelve tosses it is back to 2/3, after twenty-four it is 1/3,... so the proportion oscillates to and fro, between 2/3 and 1/3, and therefore has no well-defined limit.

Agreed, such a sequence of tosses is very unlikely, but to define 'unlikely' we need to define probability, which is what the limit is supposed to achieve. So the logic is circular. Moreover, even if the limit exists, it might not be the 'correct' value of 1/2. An extreme case occurs when the coin always lands heads. Now the limit is 1. Again, this is wildly improbable, but...

Bernoulli decided to approach the whole issue from the opposite direction. Start by simply *defining* the probability of heads and tails to be some number p between 0 and 1. Say that the coin is fair if $p = \frac{1}{2}$, and biased if not. Now Bernoulli proves a basic theorem, the law of large numbers. Introduce a reasonable rule for assigning probabilities to a series of repeated events. The law of large numbers states that in the long run, with the exception of a fraction of trials that becomes arbitrarily small, the proportion of heads does have a limit, and that limit is p. Philosophically, this theorem shows that by assigning probabilities – that is, numbers – in a natural way, the interpretation 'proportion of occurrences in the long run, ignoring rare exceptions' is valid. So Bernoulli takes the point of view that the numbers assigned as probabilities provide a consistent mathematical model of the process of tossing a coin over and over again.

His proof depends on a numerical pattern that was very familiar to Pascal. It is usually called Pascal's triangle, even though he wasn't the first person to notice it. Historians have traced its origins back to the *Chandas Shastra*, a Sanskit text attributed to Pingala, written some time between 500 BC and 200 BC. The original has not survived, but the work is known through tenth-century Hindu commentaries. Pascal's triangle looks like this:

$$
\begin{array}{ccccccccc}
 & & & & 1 & & & & \\
 & & & 1 & & 1 & & & \\
 & & 1 & & 2 & & 1 & & \\
 & 1 & & 3 & & 3 & & 1 & \\
1 & & 4 & & 6 & & 4 & & 1 \\
\end{array}
$$

where all rows start and end in 1, and each number is the sum of the two immediately above it. We now call these numbers binomial coefficients, because they arise in the algebra of the binomial (two-variable) expression $(p+q)^n$. Namely,

$$(p+q)^0 = 1$$
$$(p+q)^1 = p+q$$
$$(p+q)^2 = p^2 + 2pq + q^2$$
$$(p+q)^3 = p^3 + 3p^2q + 3pq^2 + q^3$$
$$(p+q)^4 = p^4 + 4p^3q + 6p^2q^2 + 4pq^3 + q^4$$

and Pascal's triangle is visible as the coefficients of the separate terms.

Bernoulli's key insight is that if we toss a coin n times, with a probability p of getting heads, then the probability of a specific number of tosses yielding heads is the corresponding term of $(p+q)^n$, where $q = 1 - p$. For example, suppose that I toss the coin three times. Then the eight possible results are:

HHH
HHT   HTH   THH
HTT   THT   TTH
TTT

where I've grouped the sequences according to the number of heads. So out of the eight possible sequences, there are

1 sequence with 3 heads
3 sequences with 2 heads
3 sequences with 1 heads
1 sequence with 0 heads

The link with binomial coefficients is no coincidence. If you expand the algebraic formula $(H+T)^3$ but don't collect the terms together, you get

$$HHH + HHT + HTH + THH + HTT + THT + TTH + TTT$$

Collecting terms according to the number of Hs then gives

$$H^3 + 3H^2T + 3HT^2 + T^3$$

After that, it's a matter of replacing each of H and T by its probability, p or q respectively.

Even in this case, each extreme HHH and TTT occurs only once in eight trials, and more equitable numbers occur in the other six. A more sophisticated calculation using standard properties of binomial coefficients proves Bernoulli's law of large numbers.

Advances in mathematics often come about because of ignorance. When mathematicians don't know how to calculate something important, they find a way to sneak up on it indirectly. In this case, the problem is to calculate those binomial coefficients. There's an explicit formula, but if, for instance, you want to know the probability of getting exactly 42 heads when tossing a coin 100 times, you have to do 200 multiplications and then simplify a very complicated fraction. (There are short cuts; it's still a big mess.) My computer tells me in a split second that the answer is

$$28,258,808,871,162,574,166,368,460,400\,p^{42}q^{58}$$

but Bernoulli didn't have that luxury. No one did until the 1960s, and computer algebra systems didn't really become widely available until the late 1980s.

Since this kind of direct calculation wasn't feasible, Bernoulli's immediate successors tried to find good approximations. Around 1730 Abraham De Moivre derived an approximate formula for the probabilities involved in repeated tosses of a biased coin. This led to the error function or normal distribution, often referred to as the 'bell curve' because of its shape. What he proved was this. Define the *normal distribution* $\Phi(x)$ with mean μ and variance σ^2 by the formula

$$\Phi(x) = \frac{1}{\sqrt{2\pi\sigma}} e^{-\frac{(x-\mu)^2}{2\sigma^2}}$$

Then for large n the graph of the probability of getting m heads in n tosses of a biased coin is very close to that of $\Phi(x)$ when

$$x = m/n - p \quad \mu = np \quad \sigma = npq$$

Here 'mean' refers to the average, and 'variance' is a measure of how far the data spread out – the width of the bell curve. The square root of the variance, σ itself, is called the standard deviation. Figure 32 (*left*) shows how the value of $\Phi(x)$ depends on x. The curve looks a bit like a bell, hence the informal name. The bell curve is an example of a probability distribution; this means that the probability of obtaining data between two given values is equal to the area under the curve and between the vertical lines corresponding to those values. The total area under the curve is 1, thanks to that unexpected factor $\sqrt{2\pi}$.

The idea is most easily grasped using an example. Figure 32 (*right*) shows a graph of the probabilities of getting various numbers of heads when tossing a fair coin 15 times (rectangular bars) together with the approximating bell curve.

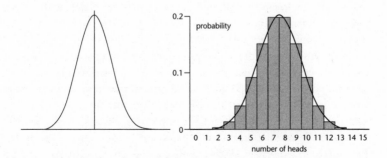

Fig 32 *Left*: Bell curve. *Right*: How it approximates the number of heads in 15 tosses of a fair coin.

The bell curve began to acquire iconic status when it started showing up in empirical data in the social sciences, not just theoretical mathematics. In 1835 Adolphe Quetelet, a Belgian who among other things pioneered quantitative methods in sociology, collected and analysed large quantities of data on crime, the divorce rate, suicide, births, deaths, human height, weight, and so on – variables that no one expected to conform to any mathematical law, because their underlying causes were too complex and involved human choices. Consider, for example, the emotional torment that drives someone to commit suicide. It seemed ridiculous to think that this could be reduced to a simple formula.

These objections make good sense if you want to predict exactly who will kill themselves, and when. But when Quetelet concentrated on statistical questions, such as the proportion of suicides in various groups of people, various locations, and different years, he started to see patterns. These proved controversial: if you predict that there will be six suicides in Paris next year, how can this make sense when each person involved has free will? They could all change their minds. But the population formed by those who do kill themselves is not specified beforehand; it comes together as a consequence of choices made not just by those who commit suicide, but by those who thought about it and didn't. People exercise free will in the context of many other things, which influence what they freely decide: here the constraints include financial problems, relationship problems, mental state, religious background... In any case, the bell curve does not make exact predictions; it just states which figure is most likely. Five or seven suicides might occur, leaving plenty of room for anyone to exercise free will and change their mind.

The data eventually won the day: for whatever reason, people *en masse* behaved more predictably than individuals. Perhaps the simplest example was height. When Quetelet plotted the proportions of people with a given height, he obtained a beautiful bell curve, Figure 33. He got the same shape of curve for many other social variables.

Quetelet was so struck by his results that he wrote a book, *Sur l'homme et le développement de ses facultés* ('Treatise on Man and the Development of His Faculties') published in 1835. In it, he introduced the notion of the 'average man', a fictitious individual who was in every respect average. It has long been noted that this doesn't entirely work: the average 'man' – that is, person, so the calculation includes males and females – has (slightly less than) one breast, one testicle, 2.3 children, and so on. Nevertheless Quetelet viewed his average man as the goal of social justice, not just a suggestive mathematical fiction. It's not quite as absurd as it sounds. For example, if

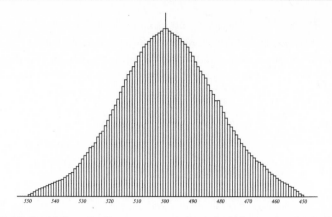

Fig 33 Quetelet's graph of how many people (vertical axis) have a given height (horizontal axis).

human wealth is spread equally to all, then everyone will have average wealth. It's not a practical goal, barring enormous social changes, but someone with strong egalitarian views might defend it as a desirable target.

The bell curve rapidly became an icon in probability theory, especially its applied arm, statistics. There were two main reasons: the bell curve was relatively simple to calculate, and there was a theoretical reason for it to occur in practice. One of the main sources for this way of thinking was eighteenth-century astronomy. Observational data are subject to errors, caused by slight variations in apparatus, human mistakes, or merely the movement of air currents in the atmosphere. Astronomers of the period wanted to observe planets, comets, and asteroids, and calculate their orbits, and this required finding whichever orbit fitted the data best. The fit would never be perfect.

The practical solution to this problem appeared first. It boiled down to this: run a straight line through the data, and choose this line so that the total error is as small as possible. Errors here have to be considered positive, and the easy way to achieve this while keeping the algebra nice is to square them. So the total error is the sum of the squares of the deviations of observations from the straight line model, and the desired line minimises this. In 1805 the French mathematician Adrien-Marie Legendre discovered a simple formula for this line, making it easy to calculate. The result is called the method of least squares. Figure 34 illustrates the method on artificial data relating stress (measured by a questionnaire) and blood pressure. The line in the figure, found using Legendre's formula, fits the data most closely according to the squared-error measure. Within ten years the method of least squares was standard among astronomers in France, Prussia, and Italy. Within another twenty years it was standard in England.

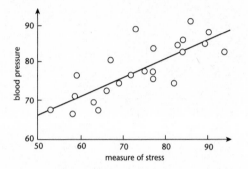

Fig 34 Using the method of least squares to relate blood pressure and stress. Dots: data. Solid line: best-fitting straight line.

Gauss made the method of least squares a cornerstone of his work in celestial mechanics. He got into the area in 1801 by successfully predicting the return of the asteroid Ceres after it was hidden in the glare of the Sun, when most astronomers thought the available data were too limited. This triumph sealed his mathematical reputation among the public and set him up for life as professor of astronomy at the University of Göttingen. Gauss didn't use least squares for this particular prediction: his calculations boiled down to solving an algebraic equation of the eighth degree, which he did by a specially invented numerical method. But in his later work, culminating in his 1809 *Theoria Motus Corporum Coelestium in Sectionibus Conicis Solem Ambientum* ('Theory of Motion of the Celestial Bodies Moving in Conic Sections around the Sun') he placed great emphasis on the method of least squares. He also stated that he had developed the idea, and used it, ten years before Legendre, which caused a bit of a fuss. It was very likely true, however, and Gauss's justification of the method was quite different. Legendre had viewed it as an exercise in curve-fitting, whereas Gauss saw it as a way to fit a probability distribution. His justification of the formula assumed that the underlying data, to which the straight line was being fitted, followed a bell curve.

It remained to justify the justification. Why should observational errors be normally distributed? In 1810 Laplace supplied an astonishing answer, also motivated by astronomy. In many branches of science it is standard to make the same observation several times, independently, and then take the average. So it is natural to model this procedure mathematically. Laplace used the Fourier transform, see Chapter 9, to prove that the average of many observations is described by a bell curve, even if the individual observations are not. His result, the central limit theorem, was a major turning point in probability and statistics, because it provided a theoretical justification for using the mathematicians' favourite distribution, the bell curve, in the analysis of observational errors.[3]

The central limit theorem singled out the bell curve as the probability distribution uniquely suited to the mean of many repeated observations. It therefore acquired the name 'normal distribution', and was seen as the default choice for a probability distribution. Not only did the normal distribution have pleasant mathematical properties, but there was also a solid reason for assuming it modelled real data. This combination of attributes proved very attractive to scientists wishing to gain insights into the social phenomena that had interested Quetelet, because it offered a way to analyse data from official records. In 1865 Francis Galton studied how a child's height relates to its parents' heights. This was part of a wider goal: understanding heredity – how human characteristics pass from parent to child. Ironically, Laplace's central limit theorem initially led Galton to doubt that this kind of inheritance existed. And, even if it did, proving that would be difficult, because the central limit theorem was a double-edged sword. Quetelet had found a beautiful bell curve for heights, but that seemed to imply very little about the different factors that affected height, because the central limit theorem predicted a normal distribution anyway, whatever the distributions of those factors might be. Even if characteristics of the parents were among those factors, they might be overwhelmed by all the others – such as nutrition, health, social status, and so on.

By 1889, however, Galton had found a way out of this dilemma. The proof of Laplace's wonderful theorem relied on averaging out the effects of many distinct factors, but these had to satisfy some stringent conditions. In 1875 Galton described these conditions as 'highly artificial', and noted that the influences being averaged

> must be (1) all independent in their effects, (2) all equal [having the same probability distribution], (3) all admitting of being treated as

simple alternatives 'above average' or 'below average'; and (4) ... calculated on the supposition that the variable influences are infinitely numerous.

None of these conditions applied to human heredity. Condition (4) corresponds to Laplace's assumption that the number of factors being added *tends to* infinity, so 'infinitely numerous' is a bit of an exaggeration; however, what the mathematics established was that to get a good approximation to a normal distribution, you had to combine a large number of factors. Each of these contributed a small amount to the average: with, say, a hundred factors, each contributed one hundredth of its value. Galton referred to such factors as 'petty'. Each on its own had no significant effect.

There was a potential way out, and Galton seized on it. The central limit theorem provided a sufficient condition for a distribution to be normal, not a necessary one. Even when its assumptions failed, the distribution concerned might still be normal *for other reasons*. Galton's task was to find out what those reasons might be. To have any hope of linking to heredity, they had to apply to a combination of a few large and disparate influences, not to a huge number of insignificant influences. He slowly groped his way towards a solution, and found it through two experiments, both dating to 1877. One was a device he called a quincunx, in which ball bearings fell down a slope, bouncing off an array of pins, with an equal chance of going left or right. In theory the balls should pile up at the bottom according to a binomial distribution, a discrete approximation to the normal distribution, so they should – and did – form a roughly bell-shaped heap, like Figure 32 (*right*). His key insight was to imagine temporarily halting the balls when they were part way down. They would still form a bell curve, but it would be narrower than the final one. Imagine releasing just one compartment of balls. It would fall to the bottom, spreading out into a tiny bell curve. The same went for any other compartment. And that meant that the final, large bell curve could be viewed as a sum of lots of tiny ones. The bell curve reproduces itself when several factors, each following its own separate bell curve, are combined.

The clincher arrived when Galton bred sweet peas. In 1875 he distributed seeds to seven friends. Each received 70 seeds, but one received very light seeds, one slightly heavier ones, and so on. In 1877 he measured the weights of the seeds of the resulting progeny. Each group was normally distributed, but the mean weight differed in each case, being comparable to the weight of each seed in the original group. When he combined the data for all of the groups, the results were again normally distributed, but the variance was bigger – the bell curve was wider. Again, this suggested that combining several bell curves led to another bell curve. Galton tracked down the mathematical reason for this. Suppose that two random variables are normally distributed, not necessarily with the same means or the same variances. Then their sum is also normally distributed; its mean is the sum of the two means, and its variance is the sum of the two variances. Obviously the same goes for sums of three, four, or more normally distributed random variables.

This theorem works when a small number of factors are combined, and each factor can be multiplied by a constant, so it actually works for any linear combination. The normal distribution is valid even when the effect of each factor is large. Now Galton could see how this result applied to heredity. Suppose that the random variable given by the height of a child is some combination of the corresponding random variables for the heights of its parents, and these are normally distributed. Assuming that the hereditary factors work by addition, the child's height will also be normally distributed.

Galton wrote his ideas up in 1889 under the title *Natural Inheritance*. In particular, he discussed an idea he called regression. When one tall parent and one short one have children, the mean height of the children should be intermediate – in fact, it should be the average of the parents' heights. The variance likewise should be the average of the variances, but the variances for the parents seemed to be roughly equal, so the variance didn't change much. As successive generations passed, the mean height would 'regress' to a fixed middle-of-the-road value, while the variance would stay pretty much unchanged. So Quetelet's neat bell curve could survive from one generation to the next. Its peak would quickly settle to a fixed value, the overall mean, while its width would stay the same. So each generation would have the same diversity of heights, despite regression to the mean. Diversity would be maintained by rare individuals who failed to regress and was self-sustaining in a sufficiently large population.

With the central role of the bell curve firmly cemented to what at the time were considered solid foundations, statisticians could build on Galton's insights and workers in other fields could apply the results. Social science was an early beneficiary, but biology soon followed, and the physical sciences were already ahead of the game thanks to Legendre, Laplace, and Gauss. Soon an entire statistical toolbox was available for anyone who wanted to extract patterns from data. I'll focus on just one technique, because it is routinely used to determine the efficacy of drugs and medical procedures, along with many other applications. It is called hypothesis testing, and its goal is to assess the significance of apparent patterns in data. It was founded by four people: the Englishmen Ronald Aylmer Fisher, Karl Pearson, and his son Egon, together with a Russian-born Pole who spent most of his life in America, Jerzy Neyman. I'll concentrate on Fisher, who developed the basic ideas when working as an agricultural statistician at Rothamstead Experimental Station, analysing new breeds of plants.

Suppose you are breeding a new variety of potato. Your data suggest that this breed is more resistant to some pest. But all such data are subject to many sources of error, so you can't be fully confident that the numbers support that conclusion – certainly not as confident as a physicist who can make very precise measurements and eliminate most errors. Fisher realised that the key issue is to distinguish a genuine difference from one arising purely by chance, and that the way to do this is to ask how probable that difference would be if only chance were involved.

Assume, for instance, that the new breed of potato appears to confer twice as much resistance, in the sense that the proportion of the new breed that survives the pest is double the proportion for the old breed. It is conceivable that this effect is due to chance, and you can calculate its probability. In fact, what you calculate is the probability of a result at least as extreme as the one observed in the data. What is the probability that the proportion of the new breed that survives the pest is at least twice what it was for the old breed? Even larger proportions are permitted here because the probability of getting *exactly* twice the proportion is bound to be very small. The wider the range of results you include, the more probable the effects of chance become, so you can have greater confidence in your conclusion if your calculation suggests it is not the result of chance. If this probability derived by this calculation is low, say 0.05, then the result is unlikely to be the result of chance; it is said to be significant at the 95% level. If the probability is lower, say 0.01, then the result is extremely unlikely to be the result of chance, and it is said to be significant at the 99% level. The percentages indicate that by chance alone, the result would not be as extreme as the one observed in 95% of trials, or in 99% of them.

Fisher described his method as a comparison between two distinct hypotheses: the hypothesis that the data are significant at the stated level, and the so-called null hypothesis that the results are due to chance. He insisted that his method must not be interpreted as confirming the hypothesis that the data are significant; it should be interpreted as a rejection of the null hypothesis. That is, it provides evidence against the data *not* being significant.

This may seem a very fine distinction, since evidence against the data not being significant surely counts as evidence in favour of it being significant. However, that's not entirely true, and the reason is that the null hypothesis has an extra built-in assumption. In order to calculate the probability that a result at least as extreme is due to chance, you need a theoretical model. The simplest way to get one is to assume a specific probability distribution. This assumption applies only in connection with the null hypothesis, because that's what you use to do the sums. You don't assume the data are normally distributed. But the default distribution for the null hypothesis is normal: the bell curve.

This built-in model has an important consequence, which 'reject the null hypothesis' tends to conceal. The null hypothesis is 'the data are due to chance'. So it is all too easy to read that statement as 'reject the data being due to chance', which in turn means you accept that they're *not* due to chance. Actually, though, the null hypothesis is 'the data are due to chance *and* the effects of chance are normally distributed', so there might be two reasons to reject the null hypothesis: the data are not due to chance, *or* they are not normally distributed. The first supports the significance of the data, but the second does not. It says you might be using the wrong statistical model.

In Fisher's agricultural work, there was generally plenty of evidence for normal distributions in the data. So the distinction I'm making didn't really matter. In other applications of hypothesis testing, though, it might. Saying that the calculations reject the null hypothesis has the virtue of being true, but because the assumption of a normal distribution is not explicitly mentioned, it is all too easy to forget that you need to check normality of the distribution of the *data* before you conclude that your results are statistically significant. As the method gets used by more and more people, who have been trained in how to do the sums but not in the assumptions behind them, there is a growing danger of wrongly assuming that the test shows your data to be significant. Especially when the normal distribution has become the automatic default assumption.

In the public consciousness, the term 'bell curve' is indelibly associated with the controversial 1994 book *The Bell Curve* by two Americans, the psychologist Richard J. Herrnstein and the political scientist Charles Murray. The main theme of the book is a claimed link between intelligence, measured by intelligence quotient (IQ), and social variables such as income, employment, pregnancy rates, and crime. The authors argue that IQ levels are better at predicting such variables than the social and economic status of the parents or their level of education. The reasons for the controversy, and the arguments involved, are complex. A quick sketch cannot really do justice to the debate, but the issues go right back to Quetelet and deserve mention.

Controversy was inevitable, no matter what the academic merits or demerits of the book might have been, because it touched a sensitive nerve: the relation between race and intelligence. Media reports tended to stress the proposal that differences in IQ have a predominantly genetic origin, but the book was more cautious about this link, leaving the interaction between genes, environment, and intelligence open. Another controversial issue was an analysis suggesting that social stratification in the United States (and indeed elsewhere) increased significantly throughout the twentieth century, and that the main cause was differences in intelligence. Yet another was a series of policy recommendations for dealing with this alleged problem. One was to reduce immigration, which the book claimed was lowering average IQ. Perhaps the most contentious was the suggestion that social welfare policies allegedly encouraging poor women to have children should be stopped.

Ironically, this idea goes back to Galton himself. His 1869 book *Hereditary Genius* built on earlier writings to develop the idea that 'a man's natural abilities are derived by inheritance, under exactly the same limitations as are the form and physical features of the whole organic world. Consequently ... it would be quite practicable to produce a highly-gifted race of men by judicious marriages during several consecutive generations.' He asserted that fertility was higher among the less intelligent, but avoided any suggestion of deliberate selection in favour of intelligence. Instead, he expressed the hope that society might change so that the more intelligent people understood the need to have plenty of children.

To many, Herrnstein and Murray's proposal to re-engineer the welfare system was uncomfortably close to the eugenics movement of the early twentieth century, in which 60,000 Americans were sterilised, allegedly because of mental illness. Eugenics became widely discredited when it became associated with Nazi Germany and the holocaust, and many of its practices are now considered to be violations of human rights legislation, in some cases amounting to crimes against humanity. Proposals to breed humans selectively are widely viewed as inherently racist. A number of social scientists endorsed the book's scientific conclusions but disputed the charge of racism; some of them were less sure about the policy proposals.

The Bell Curve initiated a lengthy debate about the methods used to compile data, the mathematical methods used to analyse them, the interpretation of the results, and the policy suggestions based on those interpretations. A task force set up by the American Psychological Association concluded that some points made in the book are valid: IQ scores are good for predicting academic achievement, this correlates with employment status, and there is no significant difference in the performance of males and females. On the other hand, the task force's report reaffirmed that both genes and environment influence IQ and it found no significant evidence that racial differences in IQ scores are genetically determined.

Other critics have argued that there are flaws in the scientific methodology, such as inconvenient data being ignored, and that the study and some responses to it may to some extent have been politically motivated. For example, it is true that social stratification has increased dramatically in the United States, but it could be argued that the main cause is the refusal of the rich to pay taxes, rather than differences in intelligence. There also seems to be an inconsistency between the alleged problem and the proposed solution. If poverty causes people to have more children, and you believe that this is a bad thing, why on earth would you want to make them even poorer?

An important part of the background, often ignored, is the definition of IQ. Rather than being something directly measurable, such as height or weight, IQ is inferred statistically from tests. Subjects are set questions, and their scores are analysed using an offshoot of the method of least squares called analysis of variance. Like the method of least squares, this technique assumes that the data are normally distributed, and it seeks to isolate those factors that determine the largest amount of variability in the data, and are therefore the most important for modelling the data. In 1904 the psychologist Charles Spearman applied this technique to several different intelligence tests. He observed that the scores that subjects obtained on different tests were highly correlated; that is, if someone did well on one test, they tended to do well on them all. Intuitively, they seemed to be measuring the same thing. Spearman's analysis showed that a single common factor – one mathematical variable, which he called g, standing for 'general intelligence' – explained almost all of the correlation. IQ is a standardised version of Spearman's g.

A key question is whether g is a real quantity or a mathematical fiction.

The answer is complicated by the methods used to choose IQ tests. These assume that the 'correct' distribution of intelligence in the population is normal – the eponymous bell curve – and calibrate the tests by manipulating scores mathematically to standardise the mean and standard deviation. A potential danger here is that you get what you expect because you take steps to filter out anything that would contradict it. Stephen Jay Gould made an extensive critique of such dangers in 1981 in *The Mismeasure of Man*, pointing out among other things that raw scores on IQ tests are often not normally distributed at all.

The main reason for thinking that *g* represents a genuine feature of human intelligence is that it is *one* factor: mathematically, it defines a single dimension. If many different tests all seem to be measuring the same thing, it is tempting to conclude that the thing concerned must be real. If not, why would the results all be so similar? Part of the answer could be that the results of IQ tests are reduced to a single numerical score. This squashes a multidimensional set of questions and potential attitudes down to a one-dimensional answer. Moreover, the test has been selected so that the score correlates strongly with the designer's view of intelligent answers – if not, no one would consider using it.

By analogy, imagine collecting data on several different aspects of 'size' in the animal kingdom. One might measure mass, another height, others length, width, diameter of left hind leg, tooth size, and so on. Each such measure would be a single number. They would in general be closely correlated: tall animals tend to weight more, have bigger teeth, thicker legs... If you ran the data through an analysis of variance you would very probably find that a single combination of those data accounted for the vast majority of the variability, just like Spearman's *g* does for different measurements of things thought to relate to intelligence. Would this necessarily imply that all of these features of animals have the same underlying cause? That *one thing* controls them all? Possibly: a growth hormone level, perhaps? But probably not. The richness of animal form does not comfortably compress into a single number. Many other features do not correlate with size at all: ability to fly, being striped or spotted, eating flesh or vegetation. The single special combination of measurements that accounts for most of the variability could be a mathematical consequence of the methods used to find it – especially if those variables were chosen, as here, to have a lot in common to begin with.

Going back to Spearman, we see that his much-vaunted *g* may be one-dimensional because IQ tests are one-dimensional. IQ is a statistical method for quantifying specific kinds of problem-solving ability, mathematically convenient but not necessarily corresponding to a real attribute of the human brain, and not necessarily representing whatever it is that we mean by 'intelligence'.

By focusing on one issue, IQ, and using that to set policy, *The Bell Curve* ignores the wider context. Even if it were sensible to genetically engineer a nation's population, why confine the process to the poor? Even if on average the poor have lower IQs than the rich, a bright poor child will outperform a dumb rich one any day, despite the obvious social and educational advantages that children of the rich enjoy. Why resort to welfare cuts when you could aim more accurately at what you claim to be the real problem: intelligence itself? Why not improve education? Indeed, why aim your policy at increasing intelligence at all? There are many other desirable human traits. Why not reduce gullibility, aggressiveness, or greed?

It is a mistake to think about a mathematical model as if it were the reality. In the physical sciences, where the model often fits reality very well, this may be a convenient way of thinking that causes little harm. But in the social sciences, models are often little better than caricatures. The choice of title for *The Bell Curve* hints at this tendency to conflate model with reality. The idea that IQ is some sort of precise measure of human ability, merely because it has a mathematical pedigree, makes the same error. It is not sensible to base sweeping and highly contentious social policy on simplistic, flawed mathematical models. The real point about *The Bell Curve*, one that it makes extensively but inadvertently, is that cleverness, intelligence, and wisdom are not the same.

Probability theory is widely used in medical trials of new drugs and treatments to test the statistical significance of data. The tests are often, but not always, based on the assumption that the underlying distribution is normal. A typical example is the detection of cancer clusters. A cluster, for some disease, is a group within which the disease occurs more frequently than expected in the overall population. The cluster may be geographical, or it may refer more metaphorically to people with a particular lifestyle, or a specific period of time. For example, retired professional wrestlers, or boys born between 1960 and 1970.

Apparent clusters may be due entirely to chance. Random numbers are seldom spread out in a roughly uniform way; instead, they often cluster together. In random simulations of the UK National Lottery, where six numbers between 1 and 49 are randomly drawn, more than half appear to show some kind of regular pattern such as two numbers being consecutive or three numbers separated by the same amount, for example 5, 9, 13. Contrary to common intuition, random is clumped. When an apparent cluster is found, the medical authorities try to assess whether it is due to chance or whether there might be some possible causal connection. At one time, most children of Israeli fighter pilots were boys. It would be easy to think of possible explanations – pilots are very virile and virile men sire more boys (not true, by the way), pilots are exposed to more radiation than normal, they experience higher g-forces – but this phenomenon was short-lived, just a random cluster. In later data it disappeared. In any population of people, it is always likely that there will be more children of one sex or the other; exact equality is very improbable. To assess the significance of the cluster, you should keep observing and see whether it persists.

However, this procrastination can't be continued indefinitely, especially if the cluster involves a serious disease. AIDS was first detected as a cluster of pneumonia cases in American homosexual men in the 1980s, for instance. Asbestos fibres as a cause of a form of lung cancer, mesothelioma, first showed up as a cluster among former asbestos workers. So statistical methods are used to assess how probable such a cluster would be if it arose for random reasons. Fisher's methods of significance testing, and related methods, are widely used for that purpose.

Probability theory is also fundamental to our understanding of risk. This word has a specific, technical meaning. It refers to the potential for some action to lead to an undesirable outcome. For example, flying in an aircraft could result in being involved in a crash, smoking cigarettes could lead to lung cancer, building a nuclear power station could lead to the release of radiation in an accident or a terrorist attack, building a dam for hydroelectric power could cause deaths if the dam collapses. 'Action' here can refer to not doing something: failing to vaccinate a child might lead to its death from a disease, for example. In this case there is also a risk associated with vaccinating the child, such as an allergic reaction. Over the whole population this risk is smaller, but for specific groups it can be larger.

Many different concepts of risk are employed in different contexts. The usual mathematical definition is that the risk associated with some action or inaction is the probability of an adverse result, multiplied by the loss that would then be incurred. By this definition a one in ten chance of killing ten people has the same level of risk as a one in a million chance of killing a million people. The mathematical definition is rational in the sense that there is a specific rationale behind it, but that doesn't mean that it is necessarily sensible. We've already seen that 'probability' refers to the long run, but for rare events the long run is very long indeed. Humans, and their societies, can adapt to repeated small numbers of deaths, but a country that suddenly lost a million people at once would be in serious trouble, because all public services and industry would simultaneously come under a severe strain. It would be little comfort to be told that over the next 10 million years, the total deaths in the two cases would be comparable. So new methods are being developed to quantify risk in such cases.

Statistical methods, derived from questions about gambling, have a huge variety of uses. They provide tools for analysing social, medical, and scientific data. Like all tools, what happens depends on how they are used. Anyone using statistical methods needs to be aware of the assumptions behind those methods, and their implications. Blindly feeding numbers into a computer and taking the results as gospel, without understanding the limitations of the method being used, is a recipe for disaster. The legitimate use of statistics, however, has improved our world out of all recognition. And it all began with Quetelet's bell curve.

8 Good vibrations
Wave Equation

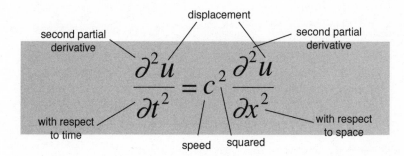

second partial
derivative

displacement

second partial
derivative

with respect
to time

speed   squared

with respect
to space

What does it say?

The acceleration of a small segment of a violin string is proportional to the average displacement of neighbouring segments.

Why is that important?

It predicts that the string will move in waves, and it generalises naturally to other physical systems in which waves occur.

What did it lead to?

Big advances in our understanding of water waves, sound waves, light waves, elastic vibrations... Seismologists use modified versions of it to deduce the structure of the interior of the Earth from how it vibrates. Oil companies use similar methods to find oil. In Chapter 11 we will see how it predicted the existence of electromagnetic waves, leading to radio, television, radar, and modern communications.

We live in a world of waves. Our ears detect waves of compression in the air: we call this 'hearing'. Our eyes detect waves of electromagnetic radiation: we call this 'seeing'. When an earthquake hits a town or a city, the destruction is caused by waves in the solid body of the Earth. When a ship bobs up and down on the ocean, it is reacting to waves in the water. Surfers use ocean waves for recreation; radio, television, and large parts of the mobile telephone network use waves of electromagnetic radiation, similar to those that we see by, but of differing wavelengths. Microwave ovens ... well, the name gives it away, doesn't it?

With so many practical instances of waves impinging on daily life, even centuries ago, the mathematicians who decided to follow up Newton's epic discovery that nature has laws could hardly fail to start thinking about waves. What got them started, though, came from the arts: specifically, music. How does a violin string create sound? What does it *do*?

There was a reason for starting with violins, the kind of reason that appeals to mathematicians, though not to governments or businessmen considering investing in mathematicians and expecting a quick payback. A violin string can sensibly be modelled as an infinitely thin line, and its motion – which is clearly the cause of the sound that the instrument makes – can be assumed to take place in a plane. This makes the problem 'low-dimensional', which means you have a chance of solving it. Once you have understood this simple example of waves, there's a good chance that the understanding can be transferred, often in small stages, to more realistic and more practical instances of waves.

The alternative, to plough headlong into highly complex problems, may appear attractive to politicians and captains of industry, but it usually gets bogged down in complexities. Mathematics thrives on simplicities, and if necessary mathematicians will invent them artificially to provide an entry route into more complex problems. They deprecatingly refer to such models as 'toys', but these are toys with a serious purpose. Toy models of waves led to today's world of electronics and high-speed global communications, wide-bodied passenger jets and artificial satellites, radio, television, tsunami warning systems... but we'd never have achieved any of those if a few mathematicians hadn't started to puzzle out how a violin works, using a model that wasn't realistic, even for a violin.

The Pythagoreans believed that the world was based on numbers, by which they meant whole numbers or ratios between whole numbers. Some of their beliefs tended towards the mystical, investing specific numbers with human attributes: 2 was male, 3 female, 5 symbolised marriage, and so on. The number 10 was very important to the Pythagoreans because it was $1+2+3+4$ and they believed there were four elements: earth, air, fire, water. This kind of speculation strikes the modern mind as slightly crazy – well, my mind, at least – but it was reasonable in an age when humans were only just starting to investigate the world around them, seeking crucial patterns. It just took a while to work out which patterns were significant and which were dross.

One of the great triumphs of the Pythagorean world view came from music. Various stories circulate: according to one, Pythagoras was passing a blacksmith's shop and he noticed that hammers of different sizes made noises of different pitch, and that hammers related by simple numbers – one twice the size of the other, for instance – made noises that harmonised. Charming though this tale is, anyone who actually tries it out with real hammers will discover that a blacksmith's operations are not especially musical, and hammers are too complicated a shape to vibrate in harmony. But there's a grain of truth: on the whole, small objects make higher-pitched noises than large ones.

The stories are on stronger ground when they refer to a series of experiments that the Pythagoreans performed using a stretched string, a rudimentary musical instrument known as a canon. We know about these experiments because Ptolemy reported them in his *Harmonics* around 150 AD. By moving a support to various positions along the string, the Pythagoreans found that when two strings of equal tension had lengths in a simple ratio, such as $2:1$ or $3:2$, they produced unusually harmonious notes. More complex ratios were discordant and unpleasant to the ear. Later scientists pushed these ideas much further, probably a bit too far: what seems pleasant to us depends on the physics of the ear, which is more complicated than that of a single string, and it also has a cultural dimension because the ears of growing children are trained by being exposed to the sounds that are common in their society. I predict that today's children will be unusually sensitive to differences in mobile phone ringtones. However, there is a solid scientific story behind these complexities, and a lot of it confirms and explains the early Pythagorean discoveries with their single-stringed experimental instrument.

Musicians describe pairs of notes in terms of the interval between them, a measure of how many steps separate them in some musical scale. The most fundamental interval is the octave, eight white notes on a piano. Notes an octave apart sound remarkably similar, except that one note is higher than the other, and they are extremely harmonious. So much so, in fact, that harmonies based on the octave can seem a bit bland. On a violin, the way to play the note one octave above an open string is to press the middle of that string against the fingerboard. A string half as long plays a note one octave higher. So the octave is associated with a simple numerical ratio of $2:1$.

Other harmonious intervals are also associated with simple numerical ratios. The most important for Western music are the fourth, a ratio of $4:3$, and the fifth, a ratio of $3:2$. The names make sense if you consider a musical scale of whole notes C D E F G A B C. With C as base, the note corresponding to a fourth is F, the fifth is G, and the octave C. If we number the notes consecutively with the base as 1, these are respectively the 4th, 5th, and 8th notes along the scale. The geometry is especially clear on an instrument like a guitar, which has segments of wire, 'frets', inserted at the relevant positions. The fret for the fourth is one-quarter of the way along the string, that for a fifth is one-third of the way along, and the octave is halfway along. You can check this with a tape measure.

These ratios provide a theoretical basis for a musical scale and led to the scale(s) now used in most Western music. The story is complex, so I'll give a simplified version. For later convenience I'll rewrite a ratio like $3:2$ as a fraction $3/2$ from now on. Start at a base note and ascend in fifths, to get strings of lengths

$$1 \quad \frac{3}{2} \quad \left(\frac{3}{2}\right)^2 \quad \left(\frac{3}{2}\right)^3 \quad \left(\frac{3}{2}\right)^4 \quad \left(\frac{3}{2}\right)^5$$

Multiplied out, these fractions become

$$1 \quad \frac{3}{2} \quad \frac{9}{4} \quad \frac{27}{8} \quad \frac{81}{16} \quad \frac{243}{32}$$

All of these notes, except the first two, are too high-pitched to remain within an octave, but we can lower them by one or more octaves, repeatedly dividing the fractions by 2 until the result lies between 1 and 2. This yields the fractions

$$1 \quad \frac{3}{2} \quad \frac{9}{8} \quad \frac{27}{16} \quad \frac{81}{64} \quad \frac{243}{128}$$

Finally, arrange these in ascending numerical order, obtaining

$$1 \quad \frac{9}{8} \quad \frac{81}{64} \quad \frac{3}{2} \quad \frac{27}{16} \quad \frac{243}{128}$$

These correspond fairly closely to the notes C D E G A B on a piano. Notice that F is missing. In fact, to the ear, the gap between 81/64 and 3/2 sounds wider than the others. To fill that gap, we insert 4/3, the ratio for the fourth, which is very close to F on the piano. It is also useful to complete the scale with a second C, one octave up, a ratio of 2. Now we obtain a musical scale based entirely on fourths, fifths, and octaves, with pitches in the ratios

$$1 \quad \frac{9}{8} \quad \frac{81}{64} \quad \frac{4}{3} \quad \frac{3}{2} \quad \frac{27}{16} \quad \frac{243}{128} \quad 2$$
$$\text{C} \quad \text{D} \quad \text{E} \quad \text{F} \quad \text{G} \quad \text{A} \quad \text{B} \quad \text{C}$$

The length is inversely proportional to the pitch, so we would have to invert the fractions to get the corresponding lengths.

We have now accounted for all the white notes on the piano, but there are also black notes. These appear because successive numbers in the scale bear two different ratios to each other: 9/8 (called a tone) and 256/243 (semitone). For example the ratio of 81/64 to 9/8 is 9/8, but that of 4/3 to 81/64 is 256/243. The names 'tone' and 'semitone' indicate an approximate comparison of the intervals. Numerically they are 1.125 and 1.05. The first is larger, so a tone corresponds to a bigger change in pitch than a semitone. Two semitones give a ratio 1.05^2, which is about 1.11; not far from 1.25. So two semitones are close to a tone. Not *very* close, I admit.

Continuing in this vein we can divide each tone into two intervals, each close to a semitone, to get a 12-note scale. This can be done in several different ways, yielding slightly different results. However it is done, there can be subtle but audible problems when changing the key of a piece of music: the intervals change slightly if, say, we move every note up a semitone. This effect could have been avoided if we had chosen a specific ratio for a semitone and arranged for its twelfth power to equal 2. Then two tones would make an exact semitone, 12 semitones would make an octave, and you could change scale by shifting all notes up or down by a fixed amount.

There is such a number, namely the twelfth root of 2, which is about 1.059, and it leads to the so-called 'equitempered scale'. It's a compromise; for example on the equitempered scale the 4/3 ratio for a fourth is $1.059^5 = 1.335$, instead of $4/3 = 1.333$. A highly trained musician can detect the difference, but it's easy to get used to it and most of us never notice.

The Pythagorean theory of harmony in nature, then, is actually built into the basis of Western music. To explain why simple ratios go hand in hand with musical harmony, we have to look at the physics of a vibrating string. The psychology of human perception also comes into the tale, but not yet.

The key is Newton's second law of motion, relating acceleration to force. You also need to know how the force exerted by a string under tension changes as the string moves, stretching or contracting slightly. For this, we use something that Newton's unwilling sparring partner Hooke discovered in 1660, called Hooke's law: the change in length of a spring is proportional to the force exerted on it. (This is not a misprint for string – a violin string is effectively a kind of spring, so the same law applies.) One obstacle remains. We can apply Newton's laws to a system composed of a finite number of masses: we get one equation per mass, and then do our best to solve the resulting system. But a violin string is a continuum, a line composed of infinitely many points. So the mathematicians of the period thought of the string as a large number of closely spaced point masses, linked together by Hooke's-law springs. They wrote down the equations, slightly simplified to make them soluble; solved them; finally they let the number of masses become arbitrarily large, and worked out what happened to the solution.

John Bernoulli carried out this programme in 1727, and the outcome was extraordinarily pretty, considering what difficulties were being swept under the carpet. To avoid confusion in the descriptions that follow, imagine that the violin is lying on its back with the string horizontal. If you pluck the string it vibrates up and down at right angles to the violin. This is the image to bear in mind. Using the bow causes the string to vibrate sideways, and the presence of the bow is confusing. In the mathematical model, all we have is one string, fixed at its ends, and no violin; the string vibrates up and down in a plane. In this set-up Bernoulli found that the shape of the vibrating string, at any instant of time, was a sine curve. The amplitude of the vibration – the maximum height of this curve – also followed a sine curve, in time rather than space. In symbols, his solution looked like sin *ct* sin *x*, where *c* is a constant, Figure 35. The spatial part sin *x* tells us the shape, but this is scaled by a factor sin *ct* at time *t*. The formula says that the string vibrates up and down, repeating the same motion over and over again. The period of oscillation, the time between successive repeats, is 2π/*c*.

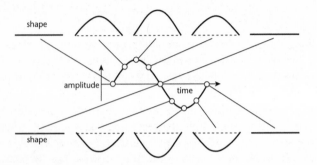

Fig 35 Successive snapshots of a vibrating string. The shape is a sine curve at each instant. The amplitude also varies sinusoidally with time.

This was the simplest solution that Bernoulli obtained, but there were others; all of them sine curves, different 'modes' of vibration, with 1, 2, 3, or more waves along the length of the string, Figure 36. Again, the sine curve was a snapshot of the shape at any instant, and its amplitude was multiplied by a time-dependent factor, which also varied sinusoidally. The formulas were sin 2*ct* sin 2*x*, sin 3*ct* sin 3*x*, and so on. The vibrational periods were 2π/2*c*, 2π/3*c*, and so on; so the more waves there were, the faster the string vibrated.

Fig 36 Snapshots of modes 1, 2, 3 of a vibrating string. In each case, the string vibrates up and down, and its amplitude varies sinusoidally with time. The more waves there are, the faster the vibration.

The string is always at rest at its ends, by the construction of the instrument and the assumptions of the mathematical model. In all modes except the first, there are additional points where the string is not vibrating; these occur where the curve crosses the horizontal axis. These 'nodes' are the mathematical reason for the occurrence of simple numerical ratios in the Pythagorean experiments. For example, since vibrational modes 2 and 3 occur in the same string, the gap between successive nodes in the mode-2 curve is 3/2 times the corresponding gap in the mode-3 curve. This explains why ratios like 3 : 2 arise naturally from the dynamics of the vibrating spring, but not why these ratios are harmonious while others are not. Before tackling this question, we introduce the main topic of this chapter: the wave equation.

The wave equation emerges from Newton's second law of motion if we apply Bernoulli's approach at the level of equations rather than solutions. In 1746 Jean Le Rond d'Alembert followed standard procedure, treating a vibrating violin string as a collection of point masses, but instead of solving the equations and looking for a pattern when the number of masses tended to infinity, he worked out what happened to the equations themselves. He derived an equation that described how the shape of the string changes over time. But before I show you what it looks like, we need a new idea, called a 'partial derivative'.

Imagine yourself in the middle of the ocean, watching waves of various shapes and sizes pass by. As they do so, you bob up and down. Physically, you can describe how your surroundings are changing in several different ways. In particular, you can focus on changes in time or changes in space. As time passes at your location, the rate at which your height changes, with respect to time, is the derivative (in the sense of calculus, Chapter 3) of your height, also with respect to time. But this doesn't describe the shape of the ocean near you, just how high the waves are as they pass you. To describe the shape, you can freeze time (conceptually) and work out how high the waves are: not just at your location, but at nearby ones. Then you can use calculus to work out how steeply the wave *slopes* at your location. Are you at a peak or trough? If so, the slope is zero. Are you halfway down the side of a wave? If so, the slope is quite large. In terms of calculus, you can put a number to that slope by working out the derivative of the wave's height with respect to space.

If a function u depends on just one variable, call it x, we write the derivative as du/dx: 'small change in u divided by small change in x'. But in the context of ocean waves the function u, the wave height, depends not just on space x but also on time t. At any fixed instant of time, we can still work out du/dx; it tells us the local slope of the wave. But instead of fixing time and letting space vary, we can also fix space and let time vary; this tells us the rate at which we are bobbing up and down. We could use the notation du/dt for this 'time derivative' and interpret it as 'small change in u divided by small change in t'. But this notation hides an ambiguity: the small change in height, du, may be, and usually is, different in the two cases. If you forget that, you are likely to get your sums wrong. When we are differentiating with respect to space, we let the space variable change a little bit and see how the height changes; when we are differentiating with respect to time, we let the time variable change a little bit and see how the height changes. There is no reason why changes over time should equal changes over space.

So mathematicians decided to remind themselves of this ambiguity by changing the symbol d to something that did not (directly) make them think 'small change'. They settled on a very cute curly d, written ∂. Then they wrote the two derivatives as $\partial u/\partial x$ and $\partial u/\partial t$. You could argue that this isn't a big advance, because it's just as easy to confuse two different meanings of ∂u. There are two answers to this criticism. One is that in this context you are not supposed to think of ∂u as a specific small change in u. The other is that using a fancy new symbol reminds you not to get confused. The second answer definitely works: as soon as you see ∂, it tells you that you will be looking at rates of change with respect to several different variables. These rates of change are called *partial derivatives*, because conceptually you change only part of the set of variables, keeping the rest fixed.

When d'Alembert worked out his equation for the vibrating string, he faced just this situation. The shape of the string depends on space – how far along the string you look – and on time. Newton's second law of motion told him that the acceleration of a small segment of string is proportional to the force that acts on it. Acceleration is a (second) time derivative. But the force is caused by neighbouring segments of the string pulling on the one we're interested in, and 'neighbouring' means small changes in *space*. When he calculated those forces, he was led to the equation

$$\frac{\partial^2 u}{\partial t^2} = c^2 \frac{\partial^2 u}{\partial x^2}$$

where $u(x,t)$ is the vertical position at location x on the string at time t, and c is a constant related to the tension in the string and how springy it is. The calculations were actually easier than Bernoulli's, because they avoided introducing special features of particular solutions.[1]

D'Alembert's elegant formula is the *wave equation*. Like Newton's second law, it is a differential equation – it involves (second) derivatives of *u*. Since these are partial derivatives, it is a *partial differential equation*. The second space derivative represents the net force acting on the string, and the second time derivative is the acceleration. The wave equation set a precedent: most of the key equations of classical mathematical physics, and a lot of the modern ones for that matter, are partial differential equations.

Once d'Alembert had written down his wave equation, he was in a position to solve it. This task was made much easier because it turned out to be a *linear* equation. Partial differential equations have many solutions, typically infinitely many, because each initial state leads to a distinct solution. For example, the violin string can in principle be bent into any shape you like before it is released and the wave equation takes over. 'Linear' means that if $u(x, t)$ and $v(x, t)$ are solutions, then so is any linear combination $au(x, t) + bv(x, t)$, where a and b are constants. Another term is 'superposition'. The linearity of the wave equation stems from the approximation that Bernoulli and d'Alembert had to make to get something they could solve: all disturbances are assumed to be small. Now the force exerted by the string can be closely approximated by a linear combination of the displacements of the individual masses. A better approximation would lead to a nonlinear partial differential equation, and life would be far more complicated. In the long run, these complications have to be tackled head-on, but the pioneers had enough to contend with already, so they worked with an approximate but very elegant equation and confined their attention to small-amplitude waves. It worked very well. In fact, it often worked pretty well for waves of larger amplitude too, a lucky bonus.

D'Alembert knew he was on the right track because he found solutions in which a fixed shape travelled along the string, just like a wave.[2] The speed of the wave turned out to be the constant c in the equation. The wave could travel either to the left or to the right, and here the superposition principle came into play. D'Alembert proved that every solution is a superposition of two waves, one travelling leftwards and the other rightwards. Moreover, each separate wave could have any shape whatsoever.[3] The standing waves found in the violin string, with fixed ends, turn out to be a combination of two waves of the same shape, one being upside down compared to the other, with one travelling to the left and the other (upside down) travelling to the right. At the ends, the two waves exactly cancel each other out: peaks of one coincide with troughs of the other. So they comply with the physical boundary conditions.

Mathematicians now had an embarrassment of riches. There were two ways to solve the wave equation: Bernoulli's, which led to sines and cosines, and d'Alembert's, which led to waves with any shape whatsoever. At first it looked as though d'Alembert's solution must be more general: sines and cosines are functions, but most functions are not sines and cosines. However, the wave equation is linear, so you could combine Bernoulli's solutions by adding constant multiples of them together. To keep it simple consider just a snapshot at a fixed time, getting rid of the time-dependence. Figure 37 shows $5 \sin x + 4 \sin 2x - 2 \cos 6x$, for example. It has a fairly irregular shape, and it wiggles a lot, but it's still smooth and wavy.

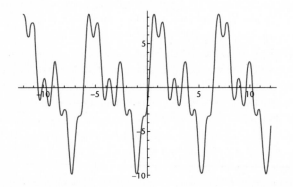

Fig 37 Typical combination of sines and cosines with various amplitudes and frequencies.

What bothered the more thoughtful mathematicians was that some functions are very rough and jagged, and you can't get those as a linear combination of sines and cosines. Well, not if you use finitely many terms – and that suggested a way out. A convergent infinite series of sines and cosines (one whose sum to infinity makes sense) also satisfies the wave equation. Does it allow jagged functions as well as smooth ones? The leading mathematicians argued about this question, which finally came to a head when the same issue turned up in the theory of heat. Problems about heat flow naturally involved discontinuous functions, with sudden jumps, which was even worse than jagged ones. I'll tell that story in Chapter 9, but the upshot is that most 'reasonable' wave shapes can be represented by an infinite series of sines and cosines, so they can be approximated as closely as you wish by finite combinations of sines and cosines.

Sines and cosines explain the harmonious ratios that so impressed the Pythagoreans. These special shapes of waves are important in the theory of sound because they represent 'pure' tones – single notes on an ideal instrument, so to speak. Any real instrument produces mixtures of pure notes. If you pluck a violin string, the main note you hear is the sin x wave, but superposed on that is a bit of sin $2x$, maybe some sin $3x$, and so on. The main note is called the fundamental and the others are its harmonics. The number in front of x is called the wave number. Bernoulli's calculations tell us that the wave number is proportional to the frequency: how many times the string vibrates, for that particular sine wave, during a single oscillation of the fundamental.

In particular, sin $2x$ has twice the frequency of sin x. What does it sound like? It is the note *one octave higher*. This is the note that sounds most harmonious when played alongside the fundamental. If you look at the shape of the string for the second mode (sin $2x$) in Figure 36, you'll notice that it crosses the axis at its midpoint as well as the two ends. At that point, a so-called node, it remains fixed. If you placed your finger at that point, the two halves of the string would still be able to vibrate in the sin $2x$ pattern, but not in the sin x one. This explains the Pythagorean discovery that a string half as long produced a note one octave higher. A similar explanation deals with the other simple ratios that they discovered: they are all associated with sine curves whose frequencies have that ratio, and such curves fit together neatly on a string of fixed length whose ends are not allowed to move.

Why do these ratios sound harmonious? Part of the explanation is that sine waves with frequencies that are not in simple ratios produce an effect called 'beats' when they are superposed. For instance, a ratio like $11:23$ corresponds to sin $11x$ + sin $23x$, which looks like Figure 38, with lots of sudden changes in shape. Another part is that the ear responds to incoming sounds in roughly the same way as the violin string. The ear, too, vibrates. When two notes beat, the corresponding sound is like a buzzing noise that repeatedly gets louder and softer. So it doesn't sound harmonious. However, there is a third part of the explanation: the ears of babies become attuned to the sounds that they hear most often. There are more nerve connections from the brain to the ear than there are in the other direction. So the brain adjusts the ear's response to incoming sounds. In other words, what we consider to be harmonious has a cultural dimension. But the simplest ratios are naturally harmonious, so most cultures use them.

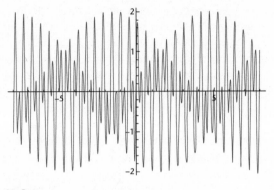

Fig 38 Beats.

Mathematicians first derived the wave equation in the simplest setting they could think of: a vibrating line, a one-dimensional system. Realistic applications required a more general theory, modelling waves in two and three dimensions. Even staying within music, a drum requires two dimensions to model the patterns in which the drumskin vibrates. The same goes for water waves on the surface of the ocean. When an earthquake strikes, the whole Earth rings like a bell, and our planet is three-dimensional. Many other areas of physics involve models with two or three dimensions. Extending the wave equation to higher dimensions turned out to be straightforward; all you had to do was repeat the same kinds of calculation that had worked for the violin string. Having learned to play the game in this simple setting, it wasn't hard to play it for real.

In three dimensions, for example, we use three space coordinates (x, y, z) and time t. The wave is described by a function u that depends on these four coordinates. For instance, this might describe the pressure in a body of air as sound waves pass through it. Making the same assumptions as d'Alembert, in particular that the amplitude of the disturbance is small, the same approach leads to an equally pretty equation:

$$\frac{\partial^2 u}{\partial t^2} = c^2 \left(\frac{\partial^2 u}{\partial x^2} + \frac{\partial^2 u}{\partial y^2} + \frac{\partial^2 u}{\partial z^2} \right)$$

The formula inside the brackets is called the Laplacian, and it corresponds to the average difference between the value of u at the point in question, and its value nearby. This expression arises so often in mathematical physics that it has its own special symbol: $\nabla^2 u$. To get the Laplacian in two dimensions, we just omit the term involving z, leading to the wave equation in that setting.

The main novelty in higher dimensions is that the shape within which the waves arise, called the domain of the equation, can be complicated. In one dimension the only connected shape is an interval, a segment of the line. In two dimensions, however, it can be any shape you can draw in the plane, and in three dimensions, any shape in space. You can model a square drum, a rectangular drum, a circular drum,[4] or a drum shaped like the silhouette of a cat. For earthquakes, you might employ a spherical domain, or for greater accuracy, an ellipsoid squashed slightly at the poles. If you are designing a car and want to eliminate unwanted vibrations, your domain should be car-shaped – or whatever part of the car the engineers want to focus on.

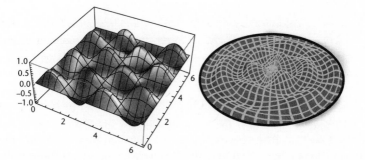

Fig 39 *Left*: Snapshot of one mode of a vibrating rectangular drum, with wave numbers 2 and 3. *Right*: Snapshot of one mode of a vibrating circular drum.

For any chosen shape of domain, there are functions analogous to Bernoulli's sines and cosines: the simplest patterns of vibration. These patterns are called modes, or normal modes if you want to make it absolutely clear what you're talking about. All other waves can be obtained by superposing normal modes, again using an infinite series if necessary. The frequencies of the normal modes represent the natural vibrational frequencies of the domain. If the domain is a rectangle, these are trigonometric functions of the form $\sin mx \cos ny$, for integers m and n, producing waves shaped like Figure 39 (*left*). If it is a circle, they are determined by new functions, called Bessel functions, with more interesting shapes, Figure 39 (*right*). The resulting mathematics applies not only to drums, but to water waves, sound waves, electromagnetic waves such as light (Chapter 11), even quantum waves (Chapter 14). It is fundamental to all of these areas. The Laplacian also turns up in equations for other physical phenomena; in particular, electric, magnetic, and gravitational fields. The mathematician's favourite trick of starting with a toy problem, one so simple that it cannot possibly be realistic, pays off big time for waves.

This is one reason why it is unwise to judge a mathematical idea by the context in which it first arises. Modelling a violin string may seem pointless when what you want to understand is earthquakes. But if you jump in at the deep end, and try to cope with all of the complexities of real earthquakes, you'll drown. You should start out paddling in the shallow end and gain confidence to swim a few lengths of the pool. Then you'll be ready for the high diving board.

The wave equation was a spectacular success, and in some areas of physics it describes reality very closely. However, its derivation requires several simplifying assumptions. When those assumptions are unrealistic, the same physical ideas can be modified to suit the context, leading to different versions of the wave equation.

Earthquakes are a typical example. Here the main problem is not d'Alembert's assumption that the amplitude of the wave is small, but changes in the physical properties of the domain. These properties can have a strong effect on seismic waves, vibrations that travel through the Earth. By understanding those effects, we can look deep inside our planet and find out what it is made of.

There are two main kinds of seismic wave: pressure waves and shear waves, usually abbreviated to P-waves and S-waves. (There are many others: this is a simplified account, covering some of the basics.) Both can occur in a solid medium, but S-waves don't occur in fluids. P-waves are waves of pressure, analogous to sound waves in air, and the changes in pressure point in the direction along which the wave propagates. Such waves are said to be longitudinal. S-waves are transverse waves, changing at right angles to the direction of travel, like the waves on a violin string. They cause solids to shear, that is, deform like a pack of cards pushed sideways, so that the cards slide along one another. Fluids don't behave like packs of cards.

When an earthquake happens, it sends out both kinds of wave. The P-waves travel faster, so a seismologist somewhere else on the Earth's surface observes those first. Then the slower S-waves arrive. In 1906 the English geologist Richard Oldham exploited this difference to make a major discovery about our planet's interior. Roughly speaking, the Earth has an iron core, surrounded by a rocky mantle, and the continents float on top of the mantle. Oldham suggested that the outer layers of the core must be liquid. If so, S-waves can't pass through those regions, but P-waves can. So there is a kind of S-wave shadow, and you can work out where it is by observing signals from earthquakes. The English mathematician Harold Jeffreys sorted out the details in 1926 and confirmed that Oldham was right.

If the earthquake is big enough, it can cause the entire planet to vibrate in one of its normal modes – the analogues for the Earth of sines and cosines for a violin. The whole planet rings like a bell, in a sense that would be literal if only we could hear the very low frequencies involved. Instruments sensitive enough to record these modes appeared in the 1960s, and they were used to observe the two most powerful earthquakes yet recorded scientifically. These were the Chilean earthquake of 1960 (magnitude 9.5) and the Alaskan earthquake of 1964 (magnitude 9.2). The first killed around 5000 people; the second killed about 130 thanks to its remote location. Both caused tsunamis and did a huge amount of damage. Both offered an unprecedented view of the Earth's deep interior, by exciting the Earth's basic vibrational modes.

Sophisticated versions of the wave equation have given seismologists the ability to see what's happening hundreds of kilometres beneath our feet. They can map the Earth's tectonic plates as one slides beneath another, known as subduction. Subduction causes earthquakes, especially so-called megathrust earthquakes like the two just mentioned. It also gives rise to mountain chains along the edges of continents, such as the Andes, and volcanoes, where the plate gets so deep that it starts to melt and magma rises to the surface. A recent discovery is that the plates need not subduct as a whole, but can break up into gigantic slabs, sinking back into the mantle at different depths.

The biggest prize in this area would be a reliable way to predict earthquakes and volcanic eruptions. This is proving elusive, because the conditions that trigger such events are complex combinations of many factors in many locations. However, some progress is being made, and the seismologists' version of the wave equation underpins many of the methods being investigated.

The same equations have more commercial applications. Oil companies prospect for liquid gold, a few kilometres underground, by setting off explosions at the surface and using returning echoes from the seismic waves they generate to map out the underlying geology. The main mathematical problem here is to reconstruct the geology from the signals received, which is a bit like using the wave equation backwards. Instead of solving the equation in a known domain to work out what the waves do, mathematicians use the observed wave patterns to reconstruct the geological features of the domain. As is often the case, working backwards like this – solving the inverse problem, in the jargon – is harder than going the other way. But practical methods exist. One of the major oil companies performs such calculations a quarter of a million times every day.

Drilling for oil has its own problems, as the blowout at the Deepwater Horizon oil rig in 2010 made clear. But at the moment, human society is heavily dependent on oil, and it would take decades to reduce this significantly, even if everyone wanted to. Next time you fill up your tank, give a thought to the mathematical pioneers who wanted to know how a violin produces its sounds. It wasn't a practical problem then, and it still isn't today. But without their discoveries, your car would take you nowhere.

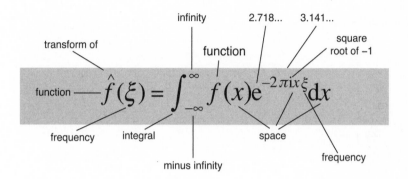

infinity   2.718...   3.141...

transform of   function   square root of −1

function

$$\hat{f}(\xi) = \int_{-\infty}^{\infty} f(x)e^{-2\pi i x \xi} dx$$

frequency   integral   space   frequency

minus infinity

What does it say?

Any pattern in space and time can be thought of as a superposition of sinusoidal patterns with different frequencies.

Why is that important?

The component frequencies can be used to analyse the patterns, create them to order, extract important features, and remove random noise.

What did it lead to?

Fourier's technique is very widely used, for example in image processing and quantum mechanics. It is used to find the structure of large biological molecules like DNA, to compress image data in digital photography, to clean up old or damaged audio recordings, and to analyse earthquakes. Modern variants are used to store fingerprint data efficiently and to improve medical scanners.

Newton's *Principia* opened the door to the mathematical study of nature, but his fellow countrymen were too obsessed with the priority dispute over calculus to find out what lay beyond. While England's finest were seething over what they perceived to be disgraceful allegations about the country's greatest living mathematician – much of it probably his own fault for listening to well-intentioned but foolish friends – their continental colleagues were extending Newton's ideas about laws of nature to most of the physical sciences. The wave equation was quickly followed by remarkably similar equations for gravitation, electrostatics, elasticity, and heat flow. Many bore the names of their inventors: Laplace's equation, Poisson's equation. The equation for heat does not; it bears the unimaginative and not entirely accurate name 'heat equation'. It was introduced by Joseph Fourier, and his ideas led to the creation of a new area of mathematics whose ramifications were to spread far beyond its original source. Those ideas could have been triggered by the wave equation, where similar methods were floating around in the collective mathematical consciousness, but history plumped for heat.

The new method had a promising beginning: in 1807 Fourier submitted an article on heat flow to the French Academy of Sciences, based on a new partial differential equation. Although that prestigious body declined to publish the work, it encouraged Fourier to develop his ideas further and try again. At that time the Academy offered an annual prize for research on whatever topic they felt was sufficiently interesting, and they made heat the topic of the 1812 prize. Fourier duly submitted his revised and extended article, and won. His heat equation looks like this:

$$\frac{\partial u}{\partial t} = \alpha \frac{\partial^2 u}{\partial x^2}$$

Here $u(x, t)$ is the temperature of a metal rod at position x and time t, considering the rod to be infinitely thin, and α is a constant, the thermal diffusivity. So it really ought to be called the temperature equation. He also developed a higher-dimensional version,

$$\frac{\partial u}{\partial t} = \alpha \nabla^2 u$$

valid on any specified region of the plane or space.

The heat equation bears an uncanny resemblance to the wave equation, with one crucial difference. The wave equation uses the second time derivative $\partial^2 u / \partial t^2$, but in the heat equation this is replaced by the first derivative $\partial u / \partial t$. This change may seem small, but its physical meaning is huge. Heat does not persist indefinitely, in the way that a vibrating violin string continues to vibrate forever (according to the wave equation, which assumes no friction or other damping). Instead, heat dissipates, dies away, as time passes, unless there is some heat source that can top it up. So a typical problem might be: heat one end of a rod to keep its temperature steady, cool the other end to do the same, and find out how the temperature varies along the rod when it settles to a steady state. The answer is that it falls off exponentially. Another typical problem is to specify the initial temperature profile along the rod, and then ask how it changes as time passes. Perhaps the left half starts at a high temperature and the right half at a cooler one; the equation then tells us how the heat from the hot part diffuses into the cooler part.

The most intriguing aspect of Fourier's prizewinning memoir was not the equation, but how he solved it. When the initial profile is a trigonometric function, such as sin x, it is easy (to those with experience in such matters) to solve the equation, and the answer is $e^{-\alpha t} \sin x$. This resembles the fundamental mode of the wave equation, but there the formula was sin ct sin x. The eternal oscillation of a violin string, corresponding to the sin ct factor, has been replaced by an exponential, and the minus sign in the exponent $-\alpha t$ tells us that the entire temperature profile dies away at the same rate, all along the rod. (The physical difference here is that waves conserve energy, but heat flow does not.) Similarly, for a profile sin $5x$, say, the solution is $e^{-25\alpha t} \sin 5x$, which also dies out, but at a much faster rate. The 25 is 5^2, and this is an example of a general pattern, applicable to initial profiles of the form sin nx or cos nx.[1] To solve the heat equation, just multiply by $e^{-n^2 \alpha t}$.

Now the story follows the same general outline as the wave equation.

The heat equation is linear, so we can superpose solutions. If the initial profile is

$$u(x,0) = \sin x + \sin 5x$$

then the solution is

$$u(x,t) = e^{-\alpha t}\sin x + e^{-25\alpha t}\sin 5x$$

and each mode dies way at a different rate. But initial profiles like this are a bit artificial. To solve the problem I mentioned earlier, we want an initial profile where $u(x, 0) = 1$ for half the rod but -1 for the other half. This profile is discontinuous, a square wave in engineering terminology. But sine and cosine curves are continuous. So no superposition of sine and cosine curves can represent a square wave.

No finite superposition, certainly. But, again, what if we allowed *infinitely many* terms? Then we can try to express the initial profile as an infinite series, of the form

$$u(x,0) = a_0 + a_1\cos x + a_2\cos 2x + a_3\cos 3x + \ldots$$
$$+ b_1\sin x + b_2\sin 2x + b_3\sin 3x + \ldots$$

for suitable constants a_0, a_1, a_2, a_3, ..., b_1, b_2, b_3, ... (There is no b_0 because $\sin 0x = 0$.) Now it does seem possible to get a square wave (see Figure 40). In fact, most coefficients can be set to zero. Only the b_n for n odd are needed, and then $b_n = 8/n\pi$.

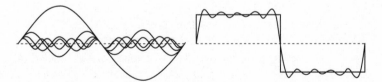

Fig 40 How to get a square wave from sines and cosines. *Left*: The component sinusoidal waves. *Right*: Their sum and a square wave. Here we show the first few terms of the Fourier series. Additional terms make the approximation to a square wave ever better.

Fourier even had general formulas for the coefficients a_n and b_n for a general profile $f(x)$, in terms of integrals:

$$a_n = \frac{1}{\pi}\int_0^{2\pi} f(x)\cos(nx)\,dx, \qquad b_n = \frac{1}{\pi}\int_0^{2\pi} f(x)\sin(nx)\,dx$$

After a lengthy trek through power series expansions of trigonometric functions, he realised that there was a much simpler way to derive these formulas. If you take two different trigonometric functions, say $\cos 2x$ and $\sin 5x$, multiply them together, and integrate from 0 to 2π, you get zero. This is even the case when they look like $\cos 5x$ and $\sin 5x$. But if they are the same – say both equal to $\sin 5x$ – the integral of their product is not zero. In fact, it is π. If you start by assuming that $f(x)$ is the sum of a trigonometric series, multiply everything by $\sin 5x$, and integrate, all of the terms disappear except for the one corresponding to $\sin 5x$, namely $b_5 \sin 5x$. Here the integral is π. Divide by that, and you have Fourier's formula for b_5. The same goes for all the other coefficients.

Although it won the academy's prize, Fourier's memoir was roundly criticised for being insufficiently rigorous, and the academy declined to publish it. This was highly unusual and it greatly irritated Fourier, but the academy held its ground. Fourier was incensed. Physical intuition told him he was right, and if you plugged his series into this equation it was clearly a solution. It *worked*. The real problem was that unwittingly he had reopened an old wound. As we saw in Chapter 8, Euler and Bernoulli had been arguing for ages about a similar issue for the wave equation, where Fourier's exponential dissipation over time was replaced by an unending sinusoidal oscillation in the wave amplitude. The underlying mathematical issues were identical. In fact, Euler had already published the integral formulas for the coefficients in the context of the wave equation.

However, Euler had never claimed that the formula worked for discontinuous functions $f(x)$, the most controversial feature of Fourier's work. The violin-string model didn't involve discontinuous initial conditions anyway – those would model a broken string, which would not vibrate at all. But for heat, it was natural to consider holding one region of a rod at one temperature and an adjacent region at a different one. In practice the transition would be smooth and very steep, but a discontinuous model was reasonable and more convenient for calculations. In fact, the solution to the heat equation explained *why* the transition would rapidly become smooth and very steep, as the heat diffused sideways. So an issue that Euler hadn't needed to worry about was becoming unavoidable, and Fourier suffered from the fallout.

Mathematicians were starting to realise that infinite series were dangerous beasts. They didn't always behave like nice finite sums. Eventually, these tangled complexities got sorted out, but it took a new view of mathematics and a hundred years of hard work to do that. In Fourier's day, everyone thought they already knew what integrals, functions, and infinite series were, but in reality it was all rather vague – 'I know one when I see one.' So when Fourier submitted his epoch-making paper, there were good reasons for the academy officials to be wary. They refused to budge, so in 1822 Fourier got round their objections by publishing his work as a book, *Théorie analytique de la chaleur* ('Analytic Theory of Heat'). In 1824 he got himself appointed secretary of the academy, thumbed his nose at all the critics, and published his original 1811 memoir, unchanged, in the academy's prestigious journal.

We now know that although Fourier was right in spirit, his critics had good reasons for worrying about rigour. The problems are subtle and the answers are not terribly intuitive. Fourier analysis, as we now call it, works very well, but it has hidden depths of which Fourier was unaware.

The question seemed to be: when does the Fourier series converge to the function it allegedly represents? That is, if you take more and more terms, does the approximation to the function get ever better? Even Fourier knew that the answer was not 'always'. It seemed to be 'usually, but with possible problems at discontinuities'. For instance at its midpoint, where the temperature jumps, the square wave's Fourier series converges – but to the wrong number. The sum is 0, but the square wave takes value 1.

For most physical purposes, it doesn't greatly matter if you change the value of a function at one isolated point. The square wave, thus modified, still *looks* square. It just does something slightly different at the discontinuity. To Fourier, this kind of issue didn't really matter. He was modelling the flow of heat, and he didn't mind if the model was a bit artificial, or needed technical changes that had no important effect on the end result. But the convergence issue could not be dismissed so lightly, because functions can have far more complicated discontinuities than a square wave.

However, Fourier was claiming that his method worked for any function, so it ought to apply even to functions such as: $f(x) = 0$ when x is rational, 1 when x is irrational. This function is discontinuous everywhere. For such functions, at that time, it wasn't even clear what the integral *meant*. And that turned out to be the real cause of the controversy. No one had defined what an integral was, not for strange functions like this one. Worse, no one had defined what a *function* was. And even if you could tidy up those omissions, it wasn't just a matter of whether the Fourier series converged. The real difficulty was to sort out *in what sense* it converged.

Resolving these issues was tricky. It required a new theory of integration, supplied by Henri Lebesgue, a reformulation of the foundations of mathematics in terms of set theory, started by Georg Cantor and opening up several entirely new cans of worms, major insights from such towering figures as Riemann, and a dose of twentieth-century abstraction to sort out the convergence issues. The final verdict was that, with the right interpretations, Fourier's idea could be made rigorous. It worked for a very broad, though not universal, class of functions. Whether the series converged to $f(x)$ for every value of x wasn't quite the right question; everything was fine provided the exceptional values of x where it didn't converge were sufficiently rare, in a precise but technical sense. If the function was continuous, the series converged for any x. At a jump discontinuity, like the change from 1 to -1 in the square wave, the series converged very democratically to the average of the values immediately to either side of the jump. But the series always converged to the function with the right interpretation of 'converge'. It converged as a whole, rather than point by point. Stating this rigorously depended on finding the right way to measure the distance between two functions. With all this in place, Fourier series did indeed solve the heat equation. But their real significance was much broader, and the main beneficiary outside pure mathematics was not the physics of heat but engineering. Especially electronic engineering.

In its most general form Fourier's method represents a signal, determined by a function f, as a combination of waves of all possible frequencies. This is called the Fourier transform of the wave. It replaces the original signal by its spectrum: a list of amplitudes and frequencies for the component sines and cosines, encoding the same information in a different way – engineers talk of transforming from the time domain to the frequency domain. When data are represented in different ways, operations that are difficult or impossible in one representation may become easy in the other. For example, you can start with a telephone conversation, form its Fourier transform, and strip out all parts of the signal whose Fourier components have frequencies too high or too low for the human ear to hear. This makes it possible to send more conversations over the same communication channels, and it's one reason why today's phone bills are, relatively speaking, so small. You can't play this game on the original, untransformed signal, because that doesn't have 'frequency' as an obvious characteristic. You don't know what to strip out.

One application of this technique is to design buildings that will survive earthquakes. The Fourier transform of the vibrations produced by a typical earthquake reveals, among other things, the frequencies at which the energy imparted by the shaking ground is greatest. A building has its own natural modes of vibration, where it will resonate with the earthquake, that is, respond unusually strongly. So the first sensible step towards earthquake-proofing a building is to make sure that the building's preferred frequencies are different from the earthquake's. The earthquake's frequencies can be obtained from observations; those of the building can be calculated using a computer model.

This is just one of many ways in which, tucked away behind the scenes, the Fourier transform affects our lives. People who live or work in buildings in earthquake zones don't need to know how to calculate a Fourier transform, but their chance of surviving an earthquake is considerably improved because some people do. The Fourier transform has become a routine tool in science and engineering; its applications include removing noise from old sound recordings, such as clicks caused by scratches on vinyl records, finding the structure of large biochemical molecules such as DNA using X-ray diffraction, improving radio reception, tidying up photographs taken from the air, sonar systems such as those used by submarines, and preventing unwanted vibrations in cars at the design stage. I'll focus here on just one of the thousands of everyday uses of Fourier's magnificent insight, one that most of us unwittingly take advantage of every time we go on holiday: digital photography.

On a recent trip to Cambodia I took about 1400 photographs, using a digital camera, and they all went on a 2 GB memory card with room for about 400 more. Now, I don't take particularly high-resolution photographs, so each photo file is about 1.1 MB. But the pictures are full colour, and they don't show any noticeable pixellation on a 27-inch computer screen, so the loss in quality isn't obvious. Somehow, my camera manages to cram into a single 2 GB card about ten times as much data as the card can possibly hold. It's like pouring a litre of milk into an eggcup. Yet it all fits in. The question is: how?

The answer is data compression. The information that specifies the image is processed to reduce its quantity. Some of this processing is 'lossless', meaning that the original raw information can if necessary be retrieved from the compressed version. This is possible because most real-world images contain redundant information. Big blocks of sky, for instance, are often the same shade of blue (well, they are where we tend to go). Instead of repeating the colour and brightness information for a blue pixel over and over again, you could store the coordinates of two opposite corners of a rectangle and a short code that means 'colour this entire region blue'. That's not quite how it's done, of course, but it shows why lossless compression is sometimes possible. When it's not, 'lossy' compression is often acceptable. The human eye is not especially sensitive to certain features of images, and these features can be recorded on a coarser scale without most of us noticing, especially if we don't have the original image to compare with. Compressing information this way is like scrambling an egg: it's easy in one direction, and does the required job, but it's not possible to reverse it. Non-redundant information is lost. It was just information that didn't do a lot to begin with, given how human vision works.

My camera, like most point-and-click ones, saves its images in files with labels like P1020339.JPG. The suffix refers to JPEG, the Joint Photographic Experts Group, and it indicates that a particular system of data compression has been used. Software for manipulating and printing photos, such as Photoshop or iPhoto, is written so that it can decode the JPEG format and turn the data back into a picture. Millions of us use JPEG files regularly, fewer are aware that they're compressed, and fewer still wonder how it's done. This is not a criticism: you don't have to know how it works to use it, that's the point. The camera and software handle it all for you. But it's often sensible to have a rough idea of what software does, and how, if only to discover how cunning some of it is. You can skip the details here if you wish: I'd like you to appreciate just how *much* mathematics goes into each image on your camera's memory card, but exactly *what* mathematics is less important.

The JPEG format[2] combines five different compression steps. The first converts the colour and brightness information, which starts out as three intensities for red, green, and blue, into three different mathematically equivalent ones that are more suited to the way the human brain perceives images. One (luminance) represents the overall brightness – what you would see with a black-and-white or 'greyscale' version of the same image. The other two (chrominance) are the differences between this and the amounts of blue and red light, respectively.

Next, the chrominance data are coarsened: reduced to a smaller range of numerical values. This step alone halves the amount of data. It does no perceptible harm because the human visual system is much less sensitive to colour differences than the camera is.

The third step uses a variant of the Fourier transform. This works not with a signal that changes over time, but with a pattern in two dimensions of space. The mathematics is virtually identical. The space concerned is an 8×8 sub-block of pixels from the image. For simplicity think just of the luminance component: the same idea applies to the colour information as well. We start with a block of 64 pixels, and for each of them we need to store one number, the luminance value for that pixel. The discrete cosine transform, a special case of the Fourier transform, decomposes the image into a superposition of standard 'striped' images instead. In half of them the stripes run horizontally; in the other half they are vertical. They are spaced at different intervals, like the various harmonics in the usual Fourier transform, and their greyscale values are a close approximation to a cosine curve. In coordinates on the block they are discrete versions of $\cos mx \cos ny$ for various integers m and n, see Figure 41.

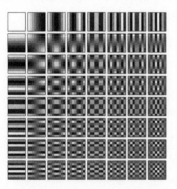

Fig 41 The 64 basic patterns from which any block of 8×8 pixels can be obtained.

This step paves the way to step four, a second exploitation of the deficiencies of human vision. We are more sensitive to variations in brightness (or colour) over large regions than we are to closely spaced variations. So the patterns in the figure can be recorded less accurately as the spacing of the stripes becomes finer. This compresses the data further. The fifth and final step uses a 'Huffman code' to express the list of strengths of the 64 basic patterns in a more efficient manner.

Every time you take a digital image using JPEG, the electronics in your camera does all of these things, except perhaps step one. (Professionals are now moving over to RAW files, which record the actual data without compression, together with the usual 'metadata' such as date, time, exposure, and so on. Files in this format take up more memory, but memory gets bigger and cheaper by the month, so that no longer matters.) A trained eye can spot the loss of image quality created by JPEG compression when the quantity of data is reduced to about 10% of the original, and an untrained eye can see it clearly by the time the file size is down to 2–3%. So your camera can record about ten times as many images on a memory card, compared with the raw image data, before anyone other than an expert would notice.

Because of applications like these, Fourier analysis has become a reflex among engineers and scientists, but for some purposes the technique has one major fault: sines and cosines go on forever. Fourier's method runs into problems when it tries to represent a compact signal. It takes huge numbers of sines and cosines to mimic a localised blip. The problem is not getting the basic shape of the blip right, but making everything outside the blip equal to zero. You have to kill off the infinitely long rippling tails of all those sines and cosines, which you do by adding on even more high-frequency sines and cosines in a desperate effort to cancel out the unwanted junk. So the Fourier transform is hopeless for blip-like signals: the transformed version is more complicated, and needs more data to describe it, than the original.

What saves the day is the generality of Fourier's method. Sines and cosines work because they satisfy one simple condition: they are mathematically independent. Formally, this means that they are *orthogonal*: in an abstract but meaningful sense, they are at right angles to each other. This is where Euler's trick, eventually rediscovered by Fourier, comes in. Multiplying two of the basic sinusoidal waveforms together and integrating over one period is a way to measure how closely related they are. If this number is large, they are very similar; if it is zero (the condition for orthogonality), they are independent. Fourier analysis works because its basic waveforms are both orthogonal and complete: they are independent and there are enough of them to represent any signal if they are suitably superposed. In effect, they provide a coordinate system on the space of all signals, just like the usual three axes of ordinary space. The main new feature is that we now have *infinitely many* axes: one for each basic waveform. But this doesn't cause many difficulties mathematically, once you get used to it. It just means you have to work with infinite series instead of finite sums, and worry a little about when the series converge.

Even in finite-dimensional spaces, there are many different coordinate systems; the axes can be rotated to point in new directions, for example. It's not surprising to find that in an infinite-dimensional space of signals, there are alternative coordinate systems that differ wildly from Fourier's. One of the most important discoveries in the whole area, in recent years, is a new coordinate system in which the basic waveforms are confined to a limited region of space. They are called wavelets, and they can represent blips very efficiently because they *are* blips.

Only recently did anyone realise that blip-like Fourier analysis was possible. Getting started is straightforward: choose a particular shape of blip, the mother wavelet (Figure 42). Then generate daughter wavelets (and granddaughters, great-granddaughters, whatever) by sliding the mother wavelet sideways into various positions, and expanding her or compressing her by a change of scale. In the same way, Fourier's basic sine and cosine curves are 'mother sinelets', and the higher-frequency sines and cosines are daughters. Being periodic, these curves cannot be blip-like.

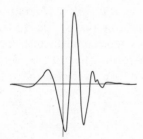

Fig 42 Daubechies wavelet.

Wavelets are designed to describe bliplike data efficiently. Moreover, because the daughter and granddaughter wavelets are just rescaled versions of mother, it is possible to focus on particular levels of detail. If you don't want to see small-scale structure, you just remove all the great-granddaughter wavelets from the wavelet transform. To represent a leopard by wavelets, you need a few big ones to get the body right, smaller ones for the eyes, nose, and of course the spots, and very tiny ones for individual hairs. To compress the data representing the leopard, you might decide that the individual hairs don't matter, so you just remove those particular component wavelets. The great thing is, the image still looks like a leopard, and it still has spots. If you try to do this with the Fourier transform of a leopard then the list of components is huge, it's not clear which items you should remove, and you probably won't recognise the result as a leopard.

All very well and good, but what shape should the mother wavelet be? For a long time nobody could work that out, or even show that a good shape exists. But in the early 1980s geophysicist Jean Morlet and mathematical physicist Alexander Grossmann found the first suitable mother wavelet. In 1985 Yves Meyer found a better mother wavelet, and in 1987 Ingrid Daubechies, a mathematician at Bell Laboratories, blew the whole field wide open. Although the previous mother wavelets looked suitably bliplike, they all had a very tiny mathematical tail that wiggled off to infinity. Daubechies found a mother wavelet with no tail at all: outside some interval, mother was always exactly zero – a genuine blip, confined entirely to a finite region of space.

The bliplike features of wavelets make them especially good for compressing images. One of their first large-scale practical uses was to store fingerprints, and the customer was the Federal Bureau of Investigation. The FBI's fingerprint database contains 300 million records, each of eight fingerprints and two thumbprints, which were originally stored as inked impressions on paper cards. This is not a convenient storage medium, so the records have been modernised by digitising the images and storing the results on a computer. Obvious advantages include being able to mount a rapid automated search for prints that match those found at the scene of a crime.

The computer file for each fingerprint card is 10 megabytes long: 80 million binary digits. So the entire archive occupies 3000 terabytes of memory: 24 quadrillion binary digits. To make matters worse, the number of new sets of fingerprints grows by 30,000 every day, so the storage requirement would grow by 2.4 trillion binary digits every day. The FBI sensibly decided that they needed some method for data compression. JPEG wasn't suitable, for various reasons, so in 2002 the FBI decided to develop a new system of compression using wavelets, the wavelet/scalar quantization (WSQ) method. WSQ reduces the data to 5% of its size by removing fine detail throughout the image. This is irrelevant to the eye's ability, as well as a computer's, to recognise the fingerprint.

There are also many recent applications of wavelets to medical imaging. Hospitals now employ several different kinds of scanner, which assemble two-dimensional cross-sections of the human body or important organs such as the brain. The techniques include CT (computerised tomography), PET (positron emission tomography), and MRI (magnetic resonance imaging). In tomography, the machine observes the total tissue density, or a similar quantity, in a single direction through the body, rather like what you would see from a fixed position if all the tissue were to become slightly transparent. A two-dimensional picture can be reconstructed by applying some clever mathematics to a whole series of such 'projections', taken at many different angles. In CT, each projection requires an X-ray exposure, so there are good reasons to limit the amount of data acquired. In all such scanning methods, less data takes less time to acquire, so more patients can use the same amount of equipment. On the other hand, good images need more data so that the reconstruction method can work more effectively. Wavelets provide a compromise, in which reducing the amount of data leads to equally acceptable images. By taking a wavelet transform, removing unwanted components, and 'detransforming' back to an image again, a poor image can be smoothed and cleaned up. Wavelets also improve the strategies by which the scanners acquire their data in the first place.

In fact, wavelets are turning up almost everywhere. Researchers in areas as wide apart as geophysics and electrical engineering are taking them on board and putting them to work in their own fields. Ronald Coifman and Victor Wickerhauser have used them to remove unwanted noise from recordings: a recent triumph was a performance of Brahms playing one of his own Hungarian Dances. It was originally recorded on a wax cylinder in 1889, which partially melted; it was re-recorded on to a 78 rpm disc. Coifman started from a radio broadcast of the disc, by which time the music was virtually inaudible amid the surrounding noise. After wavelet cleansing, you could hear what Brahms was playing – not perfectly, but at least it was audible. It's an impressive track record for an idea that first arose in the physics of heat flow 200 years ago, and was rejected for publication.

10 The ascent of humanity
Navier–Stokes Equation

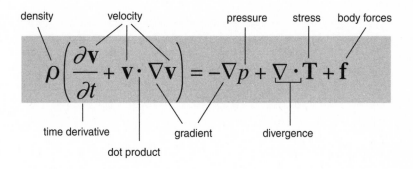

density · velocity · pressure · stress · body forces

$$\rho \left(\frac{\partial \mathbf{v}}{\partial t} + \mathbf{v} \cdot \nabla \mathbf{v} \right) = -\nabla p + \nabla \cdot \mathbf{T} + \mathbf{f}$$

time derivative · dot product · gradient · divergence

What does it say?

It's Newton's second law of motion in disguise. The left-hand side is the acceleration of a small region of fluid. The right-hand side is the forces that act on it: pressure, stress, and internal body forces.

Why is that important?

It provides a really accurate way to calculate how fluids move. This is a key feature of innumerable scientific and technological problems.

What did it lead to?

Modern passenger jets, fast and quiet submarines, Formula 1 racing cars that stay on the track at high speeds, and medical advances on blood flow in veins and arteries. Computer methods for solving the equations, known as computational fluid dynamics (CFD), are widely used by engineers to improve technology in such areas.

S een from space, the Earth is a beautiful glowing blue-and-white sphere with patches of green and brown, quite unlike any other planet in the Solar System – or any of the 500-plus planets now known to be circling other stars, for that matter. The very word 'Earth' instantly brings this image to mind. Yet a little over fifty years ago, the almost universal image for the same word would have been a handful of dirt, earth in the gardening sense. Before the twentieth century, people looked at the sky and wondered about the stars and planets, but they did so from ground level. Human flight was nothing more than a dream, the subject of myths and legends. Hardly anyone thought about travelling to another world.

A few intrepid pioneers began the slow climb into the sky. The Chinese were the first. Around 500 BC Lu Ban invented a wooden bird, which might have been a primitive glider. In 559 AD the upstart Gao Yang strapped Yuan Huangtou, the emperor's son, to a kite – against his will – to spy on the enemy from above. Yuan survived the experience but was later executed. With the seventeeth-century discovery of hydrogen the urge to fly spread to Europe, inspiring a few brave individuals to ascend into the lower reaches of Earth's atmosphere in balloons. Hydrogen is explosive, and in 1783 the French brothers Joseph-Michel and Jacques-Étienne Montgolfier gave a public demonstration of their new and much safer idea, the hot-air balloon – first with an unmanned test flight, then with Étienne as pilot.

The pace of progress, and the heights to which humans could ascend, began to increase rapidly. In 1903 Orville and Wilbur Wright made the first powered flight in an aeroplane. The first airline, DELAG (*Deutsche Luftschiffahrts-Aktiengesellschaft*), began operations in 1910, flying passengers from Frankfurt to Baden-Baden and Düsseldorf using airships made by the Zeppelin Corporation. By 1914 the St Petersburg–Tampa Airboat Line was flying passengers commercially between the two Florida cities, a journey that took 23 minutes in Tony Jannus's flying boat. Commercial air travel quickly became commonplace, and jet aircraft arrived: the De Havilland Comet began regular flights in 1952, but metal fatigue caused several crashes, and the Boeing 707 became the market leader from its launch in 1958.

Ordinary individuals could now routinely be found at an altitude of 8 kilometres, their limit to this day, at least until Virgin Galactic starts low-orbital flights. Military flights and experimental aircraft rose to greater heights. Space flight, hitherto the dream of a few visionaries, started to become a plausible proposition. In 1961 the Soviet cosmonaut Yuri Gagarin made the first manned orbit of the Earth in *Vostok 1*. In 1969 NASA's *Apollo 11* mission landed two American astronauts, Neil Armstrong and Buzz Aldrin, on the Moon. The space shuttle began operational flights in 1982, and while budget constraints prevented it achieving the original aims – a reusable vehicle with a rapid turnaround – it became one of the workhorses of low-orbit spaceflight, along with Russia's *Soyuz* spacecraft. *Atlantis* has now made the final flight of the space shuttle programme, but new vehicles are being planned, mainly by private companies. Europe, India, China, and Japan have their own space programmes and agencies.

This literal ascent of humanity has changed our view of who we are and where we live – the main reason why 'Earth' now means a blue–white globe. Those colours hold a clue to our newfound ability to fly. The blue is water, and the white is water vapour in the form of clouds. Earth is a water world, with oceans, seas, rivers, lakes. What water does best is to *flow*, often to places where it's not wanted. The flow might be rain dripping from a roof or the mighty torrent of a waterfall. It can be gentle and smooth, or rough and turbulent – the steady flow of the Nile across what would otherwise be desert, or the frothy white water of its six cataracts.

It was the patterns formed by water, or more generally any moving fluid, that attracted the attention of mathematicians in the nineteenth century, when they derived the first equations for fluid flow. The vital fluid for flight is less visible than water, but just as ubiquitous: air. The flow of air is more complex mathematically, because air can be compressed. By modifying their equations so that they applied to a compressible fluid, mathematicians initiated the science that would eventually get the Age of Flight off the ground: aerodynamics. Early pioneers might fly by rule of thumb, but commercial airliners and the space shuttle fly because engineers have done the calculations that make them safe and reliable (barring occasional accidents). Aircraft design requires a deep understanding of the mathematics of fluid flow. And the pioneer of fluid dynamics was the renowned mathematician Leonhard Euler, who died in the year the Montgolfiers made their first balloon flight.

There are few areas of mathematics towards which the prolific Euler did not turn his attention. It has been suggested that one reason for his prodigious and versatile output was politics, or more precisely, its avoidance. He worked in Russia for many years, at the court of Catherine the Great, and an effective way to avoid being caught up in political intrigue, with potentially disastrous consequences, was to be so busy with his mathematics that no one would believe he had any time to spare for politics. If this is what he was doing, we have Catherine's court to thank for many wonderful discoveries. But I'm inclined to think that Euler was prolific because he had that sort of mind. He created huge quantities of mathematics because he could do no other.

There were predecessors. Archimedes studied the stability of floating bodies over 2200 years ago. In 1738 the Dutch mathematician Daniel Bernoulli published *Hydrodynamica* ('Hydrodynamics'), containing the principle that fluids flow faster in regions where the pressure is lower. Bernoulli's principle is often invoked today to explain why aircraft can fly: the wing is shaped so that the air flows faster across the top surface, lowering the pressure and creating lift. This explanation is a bit too simplistic, and many other factors are involved in flight, but it does illustrate the close relationship between basic mathematical principles and practical aircraft design. Bernoulli embodied his principle in an algebraic equation relating velocity and pressure in an incompressible fluid.

In 1757 Euler turned his fertile mind to fluid flow, publishing an article 'Principes généraux du mouvement des fluides' (General principles of the movement of fluids) in the *Memoirs of the Berlin Academy*. It was the first serious attempt to model fluid flow using a partial differential equation. To keep the problem within reasonable bounds, Euler made some simplifying assumptions: in particular, he assumed the fluid was incompressible, like water rather than air, and had zero viscosity – no stickiness. These assumptions allowed him to find some solutions, but they also made his equations rather unrealistic. Euler's equation is still in use today for some types of problem, but on the whole it is too simple to be of much practical use.

Two scientists came up with a more realistic equation. Claude-Louis Navier was a French engineer and physicist; George Gabriel Stokes was an Irish mathematician and physicist. Navier derived a system of partial differential equations for the flow of a viscous fluid in 1822; Stokes started publishing on the topic twenty years later. The resulting model of fluid flow is now called the Navier–Stokes equation (often the plural is used because the equation is stated in terms of a vector, so it has several components). This equation is so accurate that nowadays engineers often use computer solutions instead of performing physical tests in wind tunnels. This technique, known as computational fluid dynamics (CFD), is now standard in any problem involving fluid flow: the aerodynamics of the space shuttle, the design of Formula 1 racing cars and everyday road cars, and blood circulating through the human body or an artificial heart.

There are two ways to look at the geometry of a fluid. One is to follow the movements of individual tiny particles of fluid and see where they go. The other is to focus on the velocities of such particles: how fast, and in which direction, they are moving at any instant. The two are intimately related, but the relationship is difficult to disentangle except in numerical approximations. One of the great insights of Euler, Navier, and Stokes was the realisation that everything looks a lot simpler in terms of the velocities. The flow of a fluid is best understood in terms of a velocity field: a mathematical description of how the velocity varies from point to point in space and from instant to instant in time. So Euler, Navier, and Stokes wrote down equations describing the velocity field. The actual flow patterns of the fluid can then be calculated, at least to a good approximation.

The Navier–Stokes equation looks like this:

$$\rho\left(\frac{\partial \mathbf{v}}{\partial t} + \mathbf{v} \cdot \nabla \mathbf{v}\right) = -\nabla p + \nabla \cdot \mathbf{T} + \mathbf{f}$$

where ρ is the density of the fluid, $\mathbf{v}$ is its velocity field, p is pressure, $\mathbf{T}$ determines the stresses, and $\mathbf{f}$ represents body forces – forces that act throughout the entire region, not just at its surface. The dot is an operation on vectors, and ∇ is an expression in partial derivatives, namely

$$\nabla = \left(\frac{\partial}{\partial x}, \frac{\partial}{\partial y}, \frac{\partial}{\partial z}\right)$$

The equation is derived from basic physics. As with the wave equation, a crucial first step is to apply Newton's second law of motion to relate the movement of a fluid particle to the forces that act on it. The main force is elastic stress, and this has two main constituents: frictional forces caused by the viscosity of the fluid, and the effects of pressure, either positive (compression) or negative (rarefaction). There are also body forces, which stem from the acceleration of the fluid particle itself. Combining all this information leads to the Navier–Stokes equation, which can be seen as a statement of the law of conservation of momentum in this particular context. The underlying physics is impeccable, and the model is realistic enough to include most of the significant factors; this is why it fits reality so well. Like all of the traditional equations of classical mathematical physics it is a continuum model: it assumes that the fluid is infinitely divisible.

This is perhaps the main place where the Navier–Stokes equation potentially loses touch with reality, but the discrepancy shows up only when the motion involves rapid changes on the scale of individual molecules. Such small-scale motions are important in one vital context: turbulence. If you turn on a tap and let the water flow out slowly, it arrives in a smooth trickle. Turn the tap on full, however, and you often get a surging, frothy, foaming gush of water. Similar frothy flows occur in rapids on a river. This effect is known as turbulence, and those of us who fly regularly are well aware of its effects when it occurs in air. It feels as though the aircraft is driving along a very bumpy road.

Solving the Navier–Stokes equation is hard. Until really fast computers were invented, it was so hard that mathematicians were reduced to short cuts and approximations. But when you think about what a real fluid can do, it *ought* to be hard. You only have to look at water flowing in a stream, or waves breaking on a beach, to see that fluids can flow in extremely complex ways. There are ripples and eddies, wave patterns and whirlpools, and fascinating structures like the Severn bore, a wall of water that races up the estuary of the River Severn in south-west England when the tide comes in. The patterns of fluid flow have been the source of innumerable mathematical investigations, yet one of the biggest and most basic questions in the area remains unanswered: is there a mathematical guarantee that solutions of the Navier–Stokes equation actually *exist*, valid for all future time? There is a million-dollar prize for anyone who can solve it, one of the seven Clay Institute Millennium Prize problems, chosen to represent the most important unsolved mathematical problems of our age. The answer is 'yes' in two-dimensional flow, but no one knows for three-dimensional flow.

Despite this, the Navier–Stokes equation provides a useful model of turbulent flow because molecules are extremely small. Turbulent vortices a few millimetres across already capture many of the main features of turbulence, whereas a molecule is far smaller, so a continuum model remains appropriate. The main problem that turbulence causes is practical: it makes it virtually impossible to solve the Navier–Stokes equation numerically, because a computer can't handle infinitely complex calculations. Numerical solutions of partial differential equations use a grid, dividing space into discrete regions and time into discrete intervals. To capture the vast range of scales on which turbulence operates – its big vortices, middle-sized ones, right down to the millimetre-scale ones – you need an impossibly fine computational grid. For this reason, engineers often use statistical models of turbulence instead.

The Navier–Stokes equation has revolutionised modern transport. Perhaps its greatest influence is on the design of passenger aircraft, because not only do these have to fly efficiently, but they have to *fly*, stably and reliably. Ship design also benefits from the equation, because water is a fluid. But even ordinary household cars are now designed on aerodynamic principles, not just because it makes them look sleek and cool, but because efficient fuel consumption relies on minimising drag caused by the flow of air past the vehicle. One way to reduce your carbon footprint is to drive an aerodynamically efficient car. Of course there are other ways, ranging from smaller, slower cars to electric motors, or just driving less. Some of the big improvements in fuel consumption figures have come from improved engine technology, some from better aerodynamics.

In the earliest days of aircraft design, pioneers put their aeroplanes together using back-of-the-envelope calculations, physical intuition, and trial and error. When your aim was to fly more than a hundred metres no more than three metres off the ground, that was good enough. The first time that *Wright Flyer I* got properly off the ground, instead of stalling and crashing after three seconds in the air, it travelled 120 feet at a speed just below 7 mph. Orville, the pilot on that occasion, managed to keep it aloft for a staggering 12 seconds. But the size of passenger aircraft quickly grew, for economic reasons: the more people you can carry in one flight, the more profitable it will be. Soon aircraft design had to be based on a more rational and reliable method. The science of aerodynamics was born, and its basic mathematical tools were equations for fluid flow. Since air is both viscous and compressible, the Navier–Stokes equation, or some simplification that makes sense for a given problem, took centre stage as far as theory went.

However, solving those equations, in the absence of modern computers, was virtually impossible. So the engineers resorted to an analogue computer: placing models of the aircraft in a wind tunnel. Using a few general properties of the equations to work out how variables change as the scale of the model changes, this method provided basic information quickly and reliably. Most Formula 1 teams today use wind tunnels to test their designs and evaluate potential improvements, but computer power is now so great that most also use CFD. For example, Figure 43 shows a CFD calculation of air flow past a BMW Sauber car. As I write, one team, Virgin Racing, uses only CFD, but they will be using a wind tunnel as well next year.

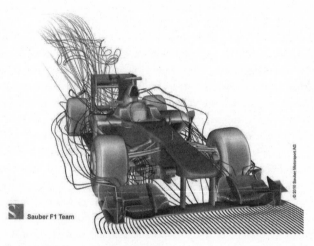

Fig 43 Computed air flow past a Formula 1 car.

Wind tunnels are not terribly convenient; they are expensive to build and run, and they need lots of scale models. Perhaps the biggest difficulty is to make accurate measurements of the flow of air without affecting it. If you put an instrument in the wind tunnel to measure, say, air pressure, then the instrument itself disturbs the flow. Perhaps the biggest practical advantage of CFD is that you can calculate the flow without affecting it. Anything you might wish to measure is easily available. Moreover, you can modify the design of the car, or a component, in software, which is a lot quicker and cheaper than making lots of different models. Modern manufacturing processes often involve computer models at the design stage anyway.

Supersonic flight, where the aircraft goes faster than sound, is especially tricky to study using models in a wind tunnel, because the wind speeds are so great. At such speeds, the air cannot move away from the aircraft as quickly as the aircraft pushes itself through the air, and this causes shockwaves – sudden discontinuities in air pressure, heard on the ground as a sonic boom. This environmental problem was one reason why the joint Anglo-French airliner Concorde, the only supersonic commercial aircraft ever to go into service, had limited success: it was not allowed to fly at supersonic speeds except over oceans. CFD is widely used to predict the flow of air past a supersonic aircraft.

There are about 600 million cars on the planet and tens of thousands of civil aircraft, so even though these applications of CFD may seem high-tech, they are significant in everyday life. Other ways to use CFD have a more human dimension. It is widely used by medical researchers to understand blood flow in the human body, for example. Heart malfunction is one of the leading causes of death in the developed world, and it can be triggered either by problems with the heart itself or by clogged arteries, which disrupt the blood flow and can cause clots. The mathematics of blood flow in the human body is especially intractable analytically because the walls of the arteries are elastic. It's difficult enough to calculate the movement of fluid through a rigid tube; it's much harder if the tube can change its shape depending on the pressure that the fluid exerts, because now the domain for the calculation doesn't stay the same as time passes. The shape of the domain affects the flow pattern of the fluid, and simultaneously the flow pattern of the fluid affects the shape of the domain. Pen-and-paper mathematics can't handle that sort of feedback loop.

CFD is ideal for this kind of problem because computers can perform billions of calculations every second. The equation has to be modified to include the effects of elastic walls, but that's mostly a matter of extracting the necessary principles from elasticity theory, another well-developed part of classical continuum mechanics. For example, a CFD calculation of how blood flows through the aorta, the main artery entering the heart, has been carried out at the École Polytechnique Féderale de Lausanne in Switzerland. The results provide information that can help doctors get a better understanding of cardiovascular problems.

They also help engineers to develop improved medical devices such as stents – small metal-mesh tubes that keep the artery open. Suncica Canic has used CFD and models of elastic properties to design better stents, deriving a mathematical theorem that caused one design to be abandoned and suggested better designs. Models of this type have become so accurate that the US Food and Drugs Administration is considering requiring any group designing stents to carry out mathematical modelling before performing clinical trials. Mathematicians and doctors are joining forces to use the Navier–Stokes equation to obtain better predictions of, and better treatments for, the main causes of heart attacks.

Another, related, application is to heart bypass operations, in which a vein is removed from elsewhere in the body and grafted into the coronary artery. The geometry of the graft has a strong effect on the blood flow. This in turn affects clotting, which is more likely if the flow has vortices because blood can become trapped in a vortex and fail to circulate properly. So here we see a direct link between the geometry of the flow and potential medical problems.

The Navier–Stokes equation has another application: climate change, otherwise known as global warming. Climate and weather are related, but different. Weather is what happens at a given place, at a given time. It may be raining in London, snowing in New York, or baking in the Sahara. Weather is notoriously unpredictable, and there are good mathematical reasons for this: see Chapter 16 on chaos. However, much of the unpredictability concerns small-scale changes, both in space and time: the fine details. If the TV weatherman predicts showers in your town tomorrow afternoon and they happen six hours later and 20 kilometres away, he thinks he did a good job and you are wildly unimpressed. Climate is the long-term 'texture' of weather – how rainfall and temperature behave when averaged over long periods, perhaps decades. Because climate averages out these discrepancies, it is paradoxically easier to predict. The difficulties are still considerable, and much of the scientific literature investigates possible sources of error, trying to improve the models.

Climate change is a politically contentious issue, despite a very strong scientific consensus that human activity over the past century or so has caused the average temperature of the Earth to rise. The increase to date sounds small, about 0.75 degrees Celsius during the twentieth century, but the climate is very sensitive to temperature changes on a global scale. They tend to make the weather more extreme, with droughts and floods becoming more common.

'Global warming' does not imply that the temperature everywhere is changing by the same tiny amount. On the contrary, there are large fluctuations from place to place and from time to time. In 2010 Britain experienced its coldest winter for 31 years, prompting the *Daily Express* to print the headline 'and *still* they claim it's global warming'. As it happens,

2010 tied with 2005 as the hottest year on record, across the globe.[1] So 'they' were right. In fact, the cold snap was caused by the jet stream changing position, pushing cold air south from the Arctic, and this happened because the Arctic was unusually *warm*. Two weeks of frost in central London does not disprove global warming. Oddly, the same newspaper reported that Easter Sunday 2011 was the hottest on record, but made no connection to global warming. On that occasion they correctly distinguished weather from climate. I'm fascinated by the selective approach.

Similarly, 'climate change' does not simply mean that the climate is changing. It has done that without human assistance repeatedly, mainly on long timescales, thanks to volcanic ash and gases, long-term variations in the Earth's orbit around the Sun, even India colliding with Asia to create the Himalayas. In the context currently under debate, 'climate change' is short for 'anthropogenic climate change' – changes in global climate caused by human activity. The main causes are the production of two gases: carbon dioxide and methane. There are greenhouse gases: they trap incoming radiation (heat) from the Sun. Basic physics implies that the more of these gases the atmosphere contains, the more heat it traps; although the planet does radiate some heat away, on balance it will get warmer. Global warming was predicted, on this basis, in the 1950s, and the predicted temperature increase is in line with what has been observed.

The evidence that carbon dioxide levels have increased dramatically comes from many sources. The most direct is ice cores. When snow falls in the polar regions, it packs together to form ice, with the most recent snow at the top and the oldest at the bottom. Air is trapped in the ice, and the conditions that prevail there leave it virtually unchanged for very long periods of time, keeping the original air in and more recent air out. With care, it is possible to measure the composition of the trapped air and to determine the date when it was trapped, very accurately. Measurements made in the Antarctic show that the concentration of carbon dioxide in the atmosphere was pretty much constant over the past 100,000 years – except for the last 200, when it shot up by 30%. The source of the excess carbon dioxide can be inferred from the proportions of carbon-13, one of the isotopes (different atomic forms) of carbon. Human activity is by far the most likely explanation.

The main reason why the skeptics have even faint glimmerings of a case is the complexity of climate forecasting. This has to be done using mathematical models, because it's about the future. No model can include every single feature of the real world, and, if it did, you could never work out what it predicted, because no computer could ever simulate it. Every discrepancy between model and reality, however insignificant, is music to the skeptics' ears. There is certainly room for differences of opinion about the likely effects of climate change, or what we should do to mitigate it. But burying our heads in the sand isn't a sensible option.

Two vital aspects of climate are the atmosphere and the oceans. Both are fluids, and both can be studied using the Navier–Stokes equation. In 2010 the UK's main science funding body, the Engineering and Physical Sciences Research Council, published a document on climate change, singling out mathematics as a unifying force: 'Researchers in meteorology, physics, geography and a host of other fields all contribute their expertise, but mathematics is the unifying language that enables this diverse group of people to implement their ideas in climate models.' The document also explained that 'The secrets of the climate system are locked away in the Navier–Stokes equation, but it is too complex to be solved directly.' Instead, climate modellers use numerical methods to calculate the fluid flow at the points of a three-dimensional grid, covering the globe from the ocean depths to the upper reaches of the atmosphere. The horizontal spacing of the grid is 100 kilometres – anything smaller would make the computations impractical. Faster computers won't help much, so the best way forward is to think harder. Mathematicians are working on more efficient ways to solve the Navier–Stokes equation numerically.

The Navier–Stokes equation is only part of the climate puzzle. Other factors include heat flow within and between the oceans and the atmosphere, the effect of clouds, non-human contributions such as volcanoes, even aircraft emissions in the stratosphere. Skeptics like to emphasise such factors to suggest the models are wrong, but most of them are known to be irrelevant. For example, every year volcanoes contribute a mere 0.6% of the carbon dioxide produced by human activity. All of the main models suggest that there is a serious problem, and humans have caused it. The main question is just how much the planet will warm up, and what level of disaster will result. Since perfect forecasting is impossible, it is in everybody's interests to make sure that our climate models are the best we can devise, so that we can take appropriate action. As the glaciers melt, the Northwest Passage opens up as Arctic ice shrinks, and Antarctic ice shelves are breaking off and sliding into the ocean, we can no longer take the risk of believing that we don't need to do anything and it will all sort itself out.

11 Waves in the ether
Maxwell's Equations

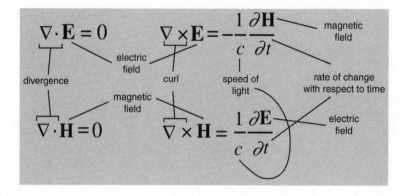

What do they say?

Electricity and magnetism can't just leak away. A spinning region of electric field creates a magnetic field at right angles to the spin. A spinning region of magnetic field creates an electric field at right angles to the spin, but in the opposite direction.

Why is that important?

It was the first major unification of physical forces, showing that electricity and magnetism are intimately interrelated.

What did it lead to?

The prediction that electromagnetic waves exist, travelling at the speed of light, so light itself is such a wave. This motivated the invention of radio, radar, television, wireless connections for computer equipment, and most modern communications.

At the start of the nineteenth century most people lit their houses using candles and lanterns. Gas lighting, which dates from 1790, was occasionally used in homes and business premises, mainly by inventors and entrepreneurs. Gas street lighting came into use in Paris in 1820. At that time, the standard way to send messages was to write a letter and send it by horse-drawn carriage; for urgent messages, keep the horse but omit the carriage. The main alternative, mostly restricted to military and official communications, was the optical telegraph. This used semaphore: mechanical devices placed on towers, which could represent letters or words in code by arranging rigid arms at various angles. These configurations could be seen through a telescope and relayed to the next tower in line. The first extensive system of this kind dates from 1792, when the French engineer Claude Chappe built 556 towers to create a 4800 kilometre network across most of France. It remained in use for sixty years.

Within a hundred years, homes and streets had electric lighting, electric telegraphy had come and gone, and people could talk to each other by telephone. Physicists had demonstrated radio communications in their laboratories, and one entrepreneur had already set up a factory selling 'wirelesses' – radio sets – to the public. Two scientists made the main discoveries that triggered this social and technological revolution. One was the Englishman Michael Faraday, who established the basic physics of electromagnetism – a tightly-knit combination of the previously separate phenomena of electricity and magnetism. The other was a Scotsman, James Clerk Maxwell, who turned Faraday's mechanical theories into mathematical equations and used them to predict the existence of radio waves travelling at the speed of light.

The Royal Institution in London is an imposing building, fronted by classical columns, tucked away on a side street near Piccadilly Circus. Today its main activity is to host popular science events for the public, but when it was founded in 1799 its brief also included 'diffusing the knowledge, and facilitating the general introduction, of useful mechanical inventions'. When John 'Mad Jack' Fuller established a Chair in Chemistry at the Royal Institution, its first incumbent was not an academic. He was the son of a would-be blacksmith, and he had trained as a bookseller's apprentice. The position allowed him to read voraciously, despite his family's lack of cash, and Jane Marcet's *Conversations on Chemistry* and Isaac Watts's *The Improvement of the Mind* inspired a deep interest in science in general and electricity in particular.

The young man was Michael Faraday. He had attended lectures at the Royal Institution given by the eminent chemist Humphry Davy, and he sent the lecturer 300 pages of notes. Shortly afterwards Davy had an accident that damaged his eyesight, and asked Faraday to become his secretary. Then an assistant at the Royal Institution got the sack, and Davy suggested Faraday as a replacement, setting him to work on the chemistry of chlorine.

The Royal Institution allowed Faraday to pursue his own scientific interests as well, and he carried out innumerable experiments on the newly discovered topic of electricity. In 1821 he learned of the work of the Danish scientist Hans Christian Ørsted, linking electricity to the much older phenomenon of magnetism. Faraday exploited this link to invent an electric motor, but Davy got upset when he didn't get any credit, and told Faraday to work on other things. Davy died in 1831, and two years later Faraday began a series of experiments on electricity and magnetism that sealed his reputation as one of the greatest scientists ever to have lived. His extensive investigations were partly motivated by the need to come up with large numbers of novel experiments to edify the man in the street and entertain the great and the good, as part of the Royal Institution's brief to encourage the public understanding of science.

Among Faraday's inventions were methods for turning electricity into magnetism and both into motion (a motor) and for turning motion into electricity (a generator). These exploited his greatest discovery, electromagnetic induction. If material that can conduct electricity moves through a magnetic field, an electrical current will flow through it. Faraday discovered this in 1831. Francesco Zantedeschi had already noticed the effect in 1829, and Joseph Henry also spotted it a little later. But Henry delayed publishing his discovery, and Faraday took the idea much further than Zantedeschi had done. Faraday's work went far beyond the Royal Institution's brief to facilitate useful mechanical inventions, by creating innovative machines that exploited frontier physics. This led, fairly directly, to electric power, lighting, and a thousand other gadgets. When others took up the baton, the whole panoply of modern electrical and electronic equipment burst upon the scene, starting with radio, moving on to television, radar, and long-distance communications. It was Faraday, more than any other single individual, who created the modern technological world, with the help of vital new ideas from hundreds of gifted engineers, scientists, and businessmen.

Being working class and lacking the normal education of a gentleman, Faraday taught himself science but not mathematics. He developed his own theories to explain and guide his experiments, but they rested on mechanical analogies and conceptual machines, not on formulas and equations. His work took its deserved place in basic physics through the intervention of one of Scotland's greatest scientific intellects, James Clerk Maxwell.

Maxwell was born the same year that Faraday announced the discovery of electromagnetic induction. One application, the electromagnetic telegraph, quickly followed, thanks to Gauss and his assistant Wilhelm Weber. Gauss wanted to use wires to carry electrical signals between Göttingen Observatory, where he hung out, to the Institute of Physics a kilometre away, where Weber worked. Presciently, Gauss simplified the previous technique for distinguishing letters of the alphabet – one wire per letter – by introducing a binary code using positive and negative current, see Chapter 15. By 1839 the Great Western Railway company was sending messages by telegraph from Paddington to West Drayton, a distance of 21 kilometres. In the same year Samuel Morse independently invented his own electric telegraph in the USA, employing Morse code (invented by his assistant Alfred Vail) and sending its first message in 1838.

In 1876, three years before Maxwell died, Alexander Graham Bell took out the first patent on a new gadget, the acoustic telegraph. It was a device that turned sound, especially speech, into electrical impulses, and transmitted them along a wire to a receiver, which turned them back into sound. We now know it as the telephone. He wasn't the first person to conceive of such a thing, or even to build one, but he held the master patent. Thomas Edison improved the design with his carbon microphone of 1878. A year later, Edison developed the carbon filament electric light bulb, and cemented himself in the popular mind as the inventor of electric lighting. In point of fact, he was preceded by at least 23 inventors, the best known being Joseph Swan, who had patented his version in 1878. In 1880, one year after Maxwell's death, the city of Wabash, Illinois became the first to use electric lighting for its streets.

These revolutions in communication and lighting owed a lot to Faraday; electrical power generation also owed a lot to Maxwell. But Maxwell's most far-reaching legacy was to make the telephone seem like a child's toy. And it stemmed, directly and inevitably, from his equations for electromagnetism.

Maxwell was born into a talented but eccentric Edinburgh family, which included lawyers, judges, musicians, politicians, poets, mining speculators, and businessmen. As a teenager he began to succumb to the charms of mathematics, winning a school competition with an essay on how to construct oval curves using pins and thread. At 16 he went to Edinburgh University, where he studied mathematics and experimented in chemistry, magnetism, and optics. He published papers in pure and applied mathematics in the Royal Society of Edinburgh's journal. In 1850 his mathematical career took a more serious turn and he moved to Cambridge University, where he was privately coached for the mathematical tripos examination by William Hopkins. The tripos in those days consisted of solving complicated problems, often involving clever tricks and extensive calculations, against the clock. Later Godfrey Harold Hardy, one of England's best mathematicians and a Cambridge professor, would have strong views about how to do creative mathematics, and cramming for a tricky examination wasn't it. In 1926 he remarked that his aim was 'not... to reform the tripos, but to destroy it'. But Maxwell crammed, and thrived, in the competitive atmosphere, probably because he had that sort of mind.

He also continued his weird experiments, among other things trying to work out how a cat always lands on its feet, even when it is held upside down only a few centimetres above a bed. The difficulty is that this appears to violate Newtonian mechanics; the cat has to rotate through 180 degrees, but has nothing to push against. The precise mechanism eluded him, and was not worked out until the French doctor Jules Marey made a series of photographs of a falling cat in 1894. The secret is that the cat is not rigid: it twists its front and back in opposite directions and back again, while extending and retracting its paws to stop these motions cancelling out.[1]

Maxwell got his mathematics degree, and continued as a postgraduate at Trinity College. There he read Faraday's *Experimental Researches* and worked on electricity and magnetism. He took up a chair of Natural Philosophy in Aberdeen, investigating Saturn's rings and the dynamics of the molecules in gases. In 1860 he moved to King's College London, and here he could sometimes meet with Faraday. Now Maxwell embarked on his most influential quest: to formulate a mathematical basis for Faraday's experiments and theories.

At the time, most physicists working on electricity and magnetism were looking for analogies with gravity. It seemed sensible: opposite electrical charges attract each other with a force which, like gravity, is proportional to the inverse square of the distance separating them. Like charges repel each other with a similarly varying force, and the same goes for magnetism, where charges are replaced by magnetic poles. The standard way of thinking was that gravity was a force whereby one body mysteriously acted on another distant body, without anything passing between the two; electricity and magnetism were assumed to act in the same manner. Faraday had a different idea: they are both 'fields', phenomena that pervade space and can be detected by the forces they produce.

What is a field? Maxwell could make little progress until he could describe the concept mathematically. But Faraday, lacking mathematical training, had posed his theories in terms of geometric structures, such as 'lines of force' along which the fields pulled and pushed. Maxwell's first great breakthrough was to reformulate these ideas by analogy with the mathematics of fluid flow, where the field in effect *is* the fluid. Lines of force were then analogous to the paths followed by the molecules of the fluid; the strength of the electric or magnetic field was analogous to the velocity of the fluid. Informally, a field was an invisible fluid; mathematically, it behaved exactly like that, whatever it really was. Maxwell borrowed ideas from the mathematics of fluids and modified them to describe magnetism. His model accounted for the main properties observed in electricity.

Not content with this initial attempt, he went on to include not just magnetism, but its relation to electricity. As the electrical fluid flowed, it affected the magnetic one, and vice versa. For magnetic fields Maxwell used the mental image of tiny vortices spinning in space. Electric fields were similarly composed of tiny charged spheres. Following this analogy and the resulting mathematics, Maxwell began to understand how a change in the electric force could create a magnetic field. As the spheres of electricity move, they cause the magnetic vortices to spin, like a football fan passing through a turnstile. The fan moves without spinning; the turnstile spins without moving.

Maxwell was slightly dissatisfied with this analogy, saying 'I do not bring it forward... as a mode of connection existing in nature... It is, however... mechanically conceivable and easily investigated, and it serves to bring out the actual mechanical connections between the known electromagnetic phenomena.' To show what he meant, he used the model to explain why parallel wires carrying opposite electrical currents repel each other, and he also explained Faraday's crucial discovery of electromagnetic induction.

The next step was to retain the mathematics while getting rid of the mechanical gadgetry that propelled the analogy. This amounted to writing down equations for the basic interactions between the electrical and magnetic fields, derived from the mechanical model, but divorced from this origin. Maxwell achieved this goal in 1864 in his famous paper 'A dynamical theory of the electromagnetic field'.

We now interpret his equations using vectors, which are quantities that possess not just a size, but a direction. The most familiar is velocity: the size is the speed, how fast the object is moving; the direction is the one along which it moves. The direction really does matter: a body moving vertically upwards at 10 kps behaves very differently from one moving vertically downwards at 10 kps. Mathematically, a vector is represented by its three components: its effect along three axes at right angles to each other, such as north/south, east/west, and up/down. The bare bones are thus that a vector is a triple (x,y,z) composed of three numbers, Figure 44. The velocity of a fluid at a given point, for instance, is a vector. In contrast, the pressure at a given point is a single number: the fancy term used to distinguish this from a vector is 'scalar'.

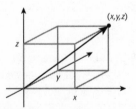

Fig 44 A three-dimensional vector.

In these terms, what is the electric field? From Faraday's perspective it is determined by lines of electrical force. In Maxwell's analogy, these are flow-lines of the electrical fluid. A flow-line tells us in which direction the fluid is flowing, and as a molecule moves along the flow-line, we can also observe its speed. For each point in space, the flow-line passing through that point therefore determines a vector, which describes the speed and direction of the electric fluid, that is, the strength and direction of the electric field *at that point*. Conversely, if we know these speeds and directions, for every point in space, we can deduce what the flow-lines look like, so in principle we know the electric field.

In short: the electric field is a system of vectors, one for each point in space. Each vector prescribes the strength and direction of the electrical force (exerted on a tiny charged test particle) at that point. Mathematicians call such a quantity a vector field: it is a function that assigns to each point in space the corresponding vector. Similarly, the magnetic field is determined by the magnetic lines of force; it is the vector field corresponding to the forces that would be exerted on a tiny magnetic test particle.

Having sorted out what electric and magnetic fields were, Maxwell could write down equations describing what they did. We now express these equations using two vector operators, known as divergence and curl. Maxwell used specific formulas involving the three components of the electric and magnetic fields. In the special case in which there are no conducting wires or metal plates, no magnets, and everything happens in a vacuum, the equations take a slightly simpler form, and I will restrict the discussion to this case.

Two of the equations tell us that the electric and magnetic fluids are incompressible – that is, electricity and magnetism cannot just leak away, they have to *go* somewhere. This translates as 'the divergence is zero', leading to the equations

$$\nabla \cdot \mathbf{E} = 0 \qquad \nabla \cdot \mathbf{H} = 0$$

where the upside-down triangle and the dot are the notation for the divergence. Two more equations tell us that when a region of electric field spins in a small circle, it creates a magnetic field at right angles to the plane of that circle, and similarly a spinning region of magnetic field creates an electric field at right angles to the plane of that circle. There is a curious twist: the electric and magnetic fields point in opposite directions for a given direction of spin. The equations are

$$\nabla \times \mathbf{E} = -\frac{1}{c}\frac{\partial \mathbf{H}}{\partial t} \qquad \nabla \times \mathbf{H} = \frac{1}{c}\frac{\partial \mathbf{E}}{\partial t}$$

where now the upside-down triangle and the cross are the notation for the curl. The symbol t stands for time, and $\partial/\partial t$ is the rate of change with respect to time. Notice that the first equation has a minus sign, but the second does not: this represents the opposite orientations that I mentioned.

What is c? It is a constant, the ratio of electromagnetic to electrostatic units. Experimentally this ratio is just under 300,000, in units of kilometres divided by seconds. Maxwell immediately recognised this number: it is the speed of light in a vacuum. Why did that quantity appear? He decided to find out. One clue, dating back to Newton, and developed by others, was the discovery that light was some kind of wave. But no one knew what the wave consisted of.

A simple calculation provided the answer. Once you know the equations for electromagnetism, you can solve them to predict how the electric and magnetic fields behave in different circumstances. You can also derive general mathematical consequences. For instance, the second pair of equations relates **E** to **H**; any mathematician will immediately try to derive equations that contain only **E** and only **H**, because that lets us concentrate on each field separately. Considering its epic consequences, this task turns out to be absurdly simple – if you have some familiarity with vector calculus. I've put the detailed working in the Notes,[2] but here's a quick summary. Following our noses, we start with the third equation, which relates the curl of **E** to the time-derivative of **H**. We don't have any other equations involving the time-derivative of **H**, but we do have one that involves the curl of **H**, namely, the fourth equation. This suggests that we should take the third equation and form the curl of both sides. Then we apply the fourth equation, simplify, and emerge with

$$\frac{\partial^2 \mathbf{E}}{\partial t^2} = c^2 \nabla^2 \mathbf{E}$$

which is the wave equation!

The same trick applied to the curl of **H** produces the same equation with **H** in place of **E**. (The minus sign is applied twice, so it disappears.) So both the electric and magnetic fields, in a vacuum, obey the wave equation. Since the same constant c occurs in each wave equation, they both travel at the same speed, namely c. So this little calculation predicts that both the electric field and the magnetic field can simultaneously support a wave – making it an electromagnetic wave, in which both fields vary in concert with each other. And the speed of that wave is ... the speed of light.

It's another of those trick questions. What travels at the speed of light? This time the answer is what you'd expect: light. But there is a momentous implication: *light is an electromagnetic wave.*

This was stupendous news. There was no reason, prior to Maxwell's derivation of his equations, to imagine such a fundamental link between light, electricity, and magnetism. But there was more. Light comes in many different colours, and once you know that light is a wave, you can work out that these correspond to waves with different wavelengths – distance betweens successive peaks. The wave equation imposes no conditions on the wavelength, so it can be anything. The wavelengths of visible light are restricted to a small range, because of the chemistry of the eye's light-detecting pigments. Physicists already knew of 'invisible light', ultraviolet and infrared. Those, of course, had wavelengths just outside the visible range. Now Maxwell's equations led to a dramatic prediction: electromagnetic waves with other wavelengths should also exist. Conceivably, any wavelength – long or short – could occur, Figure 45.

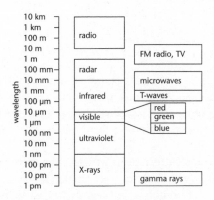

Fig 45 The electromagnetic spectrum.

No one had expected this, but as soon as theory said it ought to happen, experimentalists could go and look for it. One of them was a German, Heinrich Hertz. In 1886 he constructed a device that could generate radio waves and another that could receive them. The transmitter was little more than a machine that could produce a high-voltage spark; theory indicated that such a spark would emit radio waves. The receiver was a circular loop of copper wire, whose size was chosen to resonate with the incoming waves. A small gap in the loop, a few hundredths of a millimetre across, would reveal those waves by producing tiny sparks. In

1887 Hertz did the experiment, and it was a success. He went on to investigate many different features of radio waves. He also measured their speed, getting an answer close to the speed of light, which confirmed Maxwell's prediction and confirmed that his apparatus really was detecting electromagnetic waves.

Hertz knew that his work was important as physics, and he published it in *Electric Waves: being researches on the propagation of electric action with finite velocity through space*. But it never occurred to him that the idea might have practical uses. When asked, he replied 'It's of no use whatsoever... just an experiment that proves Maestro Maxwell was right – we just have these mysterious electromagnetic waves that we cannot see with the naked eye. But they are there.' Pressed for his view of the implications, he said 'Nothing, I guess.'

Was it a failure of imagination, or just a lack of interest? It's hard to tell. But Hertz's 'useless' experiment, confirming Maxwell's prediction of electromagnetic radiation, would quickly lead to an invention that made the telephone look like a children's toy.

Radio.

Radio makes use of an especially intriguing range of the spectrum: waves with wavelengths much longer than light. Such waves would be likely to retain their structure over long distances. The key idea, the one that Hertz missed, is simple: if you could somehow impress a signal on a wave of that kind, you could talk to the world.

Other physicists, engineers, and entrepreneurs were more imaginative, and quickly spotted radio's potential. To realise that potential, however, they had to solve a number of technical problems. They needed a transmitter that could produce a sufficiently powerful signal, and something to receive it. Hertz's apparatus was restricted to a distance of a few feet; you can understand why he didn't suggest communication as a possible application. Another problem was how to impose a signal. A third was how far the signal could be sent, which might well be limited by the curvature of the Earth. If a straight line between transmitter and receiver hits the ground, this would presumably block the signal. Later it turned out that nature has been kind to us, and the Earth's ionosphere reflects radio waves in a wide range of wavelengths, but before this was discovered there were obvious ways round the potential problem anyway. You could build tall towers and put the transmitters and receivers on those. By relaying signals from one tower to another, you could send messages round the globe, very fast.

There are two relatively obvious ways to impress a signal on a radio wave. You can make the amplitude vary or you can make the frequency vary. These methods are called amplitude-modulation and frequency-modulation: AM and FM. Both were used and both still exist. That was one problem solved. By 1893 the Serbian engineer Nikola Tesla had invented and built all of the main devices needed for radio transmission, and he had demonstrated his methods to the public. In 1894 Oliver Lodge and Alexander Muirhead sent a radio signal from the Clarendon laboratory in Oxford to a nearby lecture theatre. A year later the Italian inventor Guglielmo Marconi transmitted signals over a distance of 1.5 kilometres using new apparatus he had invented. The Italian government declined to finance further work, so Marconi moved to England. With the support of the British Post Office he soon improved the range to 16 kilometres. Further experiments led to Marconi's law: the distance over which signals can be sent is roughly proportional to the square of the height of the transmitting antenna. Make the tower twice as tall and the signal goes four times as far. This, too, was good news: it suggested that long-range transmission should be practical. Marconi set up a transmitting station on the Isle of Wight in the UK in 1897, and opened a factory the next year, making what he called 'wirelesses'. We still called them that in 1952, when I listened to the Goon Show and Dan Dare on the wireless in my bedroom, but even then we also referred to the device as 'the radio'. The word 'wireless' has of course come back into vogue, but now it is the links between your computer and its keyboard, mouse, modem, and Internet router that are wireless, rather than the link from your receiver to a distant transmitter. It's still done by radio.

Initially Marconi owned the main patents to radio, but he lost them to Tesla in 1943 in a court battle. Technological advances quickly made those patents obsolete. From 1906 to the 1950s, the vital electronic component of a radio was the vacuum tube, like a smallish light bulb, so radios had to be big and bulky. The transistor, a much smaller and more robust device, was invented in 1947 at Bell Laboratories by an engineering team that included William Shockley, Walter Brattain, and John Bardeen (see Chapter 14). By 1954 transistor radios were on the market, but radio was already losing its primacy as an entertainment medium.

By 1953, I'd already seen the future. It was the coronation of Queen Elizabeth II, and my aunt in Tonbridge had ... *a television set!* So we piled into my father's rickety car and drove 40 miles to watch the event. I was more impressed by Bill and Ben the Flowerpot Men than by the coronation, to be honest, but from that moment radio was no longer the epitome of modern household entertainment. Soon we, too, possessed a television set. Anyone who has grown up with 48-inch flatscreen colour TVs with high definition and a thousand channels will be appalled to hear that in those days the picture was black-and-white, about 12 inches across, and (in the UK) there was exactly one channel, the BBC. When we watched 'the television' it really meant *the* television.

Entertainment was just one application of radio waves. They were also vital to the military, for communications and other purposes. The invention of radar (radio detection and ranging) may well have won World War II for the Allies. This top-secret device made it possible to detect aircraft, especially enemy aircraft, by bouncing radio signals off them and observing the reflected waves. The urban myth that carrots are good for your eyesight originated in wartime disinformation, intended to stop the Nazis wondering why the British were getting so good at spotting raiding bombers. Radar has peacetime uses as well. It is how air traffic controllers keep tabs on where all the planes are, to prevent collisions; it guides passenger jets to the runway in fog; it warns pilots of imminent turbulence. Archaeologists use ground-penetrating radar to locate likely sites for the remains of tombs and ancient structures.

X-rays, first studied systematically by Wilhelm Röntgen in 1875, have much shorter wavelengths than light. This makes them more energetic, so they can pass through opaque objects, notably the human body. Doctors could use X-rays to detect broken bones and other physiological problems, and still do, although modern methods are more sophisticated and subject the patient to far less damaging radiation. X-ray scanners can now create three-dimensional images of a human body, or some part of it, in a computer. Other kinds of scanner can do the same thing using different physics.

Microwaves are efficient ways to send telephone signals, and they also turn up in the kitchen in microwave ovens, quick ways to heat food. One of the latest applications to emerge is in airport security. Terahertz radiation, otherwise known as T-waves, can penetrate clothing and even body cavities. Customs officials can use them to spot drug smugglers and terrorists. Their use is a little controversial, since they amount to an electronic strip-search, but most of us seem to think that's a small price to pay if it stops a plane being blown up or cocaine hitting the streets. T-waves are also useful to art historians, because they can reveal murals covered in layers of plaster. Manufacturers and commercial carriers can use T-waves to inspect products without taking them out of their boxes.

The electromagnetic spectrum is so versatile, and so effective, that its influence is now felt in virtually all spheres of human activity. It makes things possible that to any previous generation would appear miraculous. It took a vast number of people, from every profession, to turn the possibilities inherent in the mathematical equations into real gadgets and commercial systems. But none of this was possible until someone realised that electricity and magnetism can join forces to create a wave. The whole panoply of modern communications, from radio and television to radar and microwave links for mobile phones, was then inevitable. And it all stemmed from four equations and a couple of lines of basic vector calculus.

Maxwell's equations didn't just change the world. They opened up a new one.

12 Law and disorder
Second Law of Thermodynamics

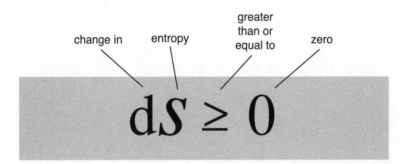

What does it say?

The amount of disorder in a thermodynamic system always increases.

Why is that important?

It places limits on how much useful work can be extracted from heat.

What did it lead to?

Better steam engines, estimates of the efficiency of renewable energy, the 'heat death of the universe' scenario, proof that matter is made of atoms, and paradoxical connections with the arrow of time.

n May 1959 the physicist and novelist C.P. Snow delivered a lecture with the title *The Two Cultures*, which provoked widespread controversy. The response of the prominent literary critic F.R. Leavis was typical of the other side of the argument; he said bluntly that there was only *one* culture: his. Snow suggested that the sciences and the humanities had lost touch with each other, and argued that this was making it very difficult to solve the world's problems. We see the same today with climate change denial and attacks on evolution. The motivation may be different, but cultural barriers help such nonsense to thrive – though it is politics that drives it.

Snow was particularly unhappy about what he saw as declining standards of education, saying:

A good many times I have been present at gatherings of people who, by the standards of the traditional culture, are thought highly educated and who have with considerable gusto been expressing their incredulity at the illiteracy of scientists. Once or twice I have been provoked and have asked the company how many of them could describe the Second Law of Thermodynamics, the law of entropy. The response was cold: it was also negative. Yet I was asking something which is about the scientific equivalent of: 'Have you read a work of Shakespeare's?'

Perhaps he sensed he was asking too much – many qualified scientists can't state the second law of thermodynamics. So he later added:

I now believe that if I had asked an even simpler question – such as, What do you mean by mass, or acceleration, which is the scientific equivalent of saying, 'Can you read?' – not more than one in ten of the highly educated would have felt that I was speaking the same language. So the great edifice of modern physics goes up, and the majority of the cleverest people in the western world have about as much insight into it as their Neolithic ancestors would have had.

Taking Snow literally, my aim in this chapter is to take us out of the Neolithic age. The word 'thermodynamics' contains a clue: it appears to mean the dynamics of heat. Can heat be dynamic? Yes: heat can *flow*. It can move from one location to another, from one object to another. Go outside on a winter's day and you soon feel cold. Fourier had written down the first serious model of heat flow, Chapter 9 and done some beautiful mathematics. But the main reason scientists were becoming interested in heat flow was a newfangled and highly profitable item of technology: the steam engine.

There is an oft-repeated story of James Watt as a boy, sitting in his mother's kitchen watching boiling steam lift the lid off a kettle, and his sudden flash of inspiration: *steam can perform work*. So, when he grew up, he invented the steam engine. It's inspirational stuff, but like many such tales this one is just hot air. Watt didn't invent the steam engine, and he didn't learn about the power of steam until he was an adult. The story's conclusion about the power of steam is true, but even in Watt's day it was old hat.

Around 50 BC the Roman architect and engineer Vitruvius described a machine called an aeolipile in his *De Architectura* ('On Architecture'), and the Greek mathematician and engineer Hero of Alexandria built one a century later. It was a hollow sphere with some water inside, and two tubes poked out, bent at an angle as in Figure 46. Heat the sphere and the water turns to steam, escapes through the ends of the tubes, and the reaction makes the sphere spin. It was the first steam engine, and it proved that steam could do work, but Hero did nothing with it beyond entertaining

Fig 46 Hero's aeolipile.

people. He did make a similar machine using hot air in an enclosed chamber to pull a rope that opened the doors of a temple. This machine had a practical application, producing a religious miracle, but it wasn't a steam engine.

Watt learned that steam could be a source of power in 1762 when he was 26 years old. He didn't discover it watching a kettle: his friend John Robison, a professor of natural philosophy at the University of Edinburgh, told him about it. But practical steam power was much older. Its discovery is often credited to the Italian engineer and architect Giovanni Branca, whose *Le Machine* ('Machine') of 1629 contained 63 woodcuts of mechanical gadgets. One shows a paddlewheel that would spin on its axle when steam from a pipe collided with its vanes. Branca speculated that this machine might be useful for grinding flour, lifting water, and cutting up wood, but it was probably never built. It was more of a thought experiment, a mechanical pipedream like Leonardo da Vinci's flying machine.

In any case, Branca was anticipated by Taqi al-Din Muhammad ibn Ma'ruf al-Shami al-Asadi, who lived around 1550 in the Ottoman Empire and was widely held to be the greatest scientist of his age. His achievements are impressive. He worked in everything from astrology to zoology, including clock-making, medicine, philosophy, and theology, and he wrote over 90 books. In his 1551 *Al-turuq al-samiyya fi al-alat al-ruhaniyya* ('The Sublime Methods of Spiritual Machines'), al-Din described a primitive steam turbine, saying that it could be used to turn roasting meat on a spit.

The first truly practical steam engine was a water pump invented by Thomas Savery in 1698. The first to make commercial profits, built by Thomas Newcomen in 1712, triggered the Industrial Revolution. But Newcomen's engine was very inefficient. Watt's contribution was to introduce a separate condenser for the steam, reducing heat loss. Developed using money provided by the entrepreneur Matthew Bolton, this new type of engine used only a quarter as much coal, leading to huge savings. Boulton and Watt's machine went into production in 1775, more than 220 years after al-Din's book. By 1776, three were up and running: one in a coal mine at Tipton, one in a Shropshire ironworks, and one in London.

Steam engines performed a variety of industrial tasks, but by far the commonest was pumping water from mines. It cost a lot of money to develop a mine, but as the upper layers became worked out and operators were forced to dig deeper into the ground, they hit the water table. It was worth spending quite a lot of money to pump the water out, since the alternative was to close the mine and start again somewhere else – and that might not even be feasible. But no one wanted to pay more than they had to, so a manufacturer who could design and build a more efficient steam engine would corner the market. So the basic question of how efficient a steam engine could be cried out for attention. Its answer did more than just describe the limits to steam engines: it created a new branch of physics, whose applications were almost boundless. The new physics shed light on everything from gases to the structure of the entire universe; it applied not just to the dead matter of physics and chemistry, but perhaps also to the complex processes of life itself. It was called thermodynamics: the motion of heat. And, just as the law of conservation of energy in mechanics ruled out mechanical perpetual motion machines, the laws of thermodynamics ruled out similar machines using heat.

One of those laws, the first law of thermodynamics, revealed a new form of energy associated with heat, and extended the law of conservation of energy (Chapter 3) into the new realm of heat engines. Another, without any previous precedent, showed that some potential ways to exchange heat, which did not conflict with conservation of energy, were nevertheless impossible because they would have to create order from disorder. This was the second law of thermodynamics.

Thermodynamics is the mathematical physics of gases. It explains how large-scale features like temperature and pressure arise from the way the gas molecules interact. The subject began with a series of laws of nature relating temperature, pressure, and volume. This version is called classical thermodynamics, and did not involve molecules – at that time few scientists believed in them. Later, the gas laws were underpinned by a further layer of explanation, based on a simple mathematical model explicitly involving molecules. The gas molecules were thought of as tiny spheres that bounced off each other like perfectly elastic billiard balls, with no energy being lost in the collision. Although molecules are not spherical, this model proved to be remarkably effective. It is called the kinetic theory of gases, and it led to experimental proof that molecules exist.

The early gas laws emerged in fits and starts over a period of nearly fifty years, and are mainly attributed to the Irish physicist and chemist Robert Boyle, the French mathematician and balloon pioneer Jacques Alexandre César Charles, and the French physicist and chemist Joseph Louis Gay-Lussac. However, many of the discoveries were made by others. In 1834, the French engineer and physicist Émile Clapeyron combined all of these laws into one, the ideal gas law, which we now write as

$$pV = RT$$

Here p is pressure, V is volume, T the temperature, and R is a constant. The equation states that pressure times volume is proportional to temperature. It took a lot of work with many different gases to confirm each separate law, and Clapeyron's overall synthesis, experimentally. The word 'ideal' appears because real gases do not obey the law in all circumstances, especially at high pressures where interatomic forces come into play. But the ideal version was good enough for designing steam engines.

Thermodynamics is encapsulated in a number of more general laws, not reliant on the precise form of the gas law. However, it does require there to be some such law, because temperature, pressure, and volume are not independent. There has to be some relation between them, but it doesn't greatly matter what.

The first law of thermodynamics stems from the mechanical law of conservation of energy. In Chapter 3 we saw that there are two distinct kinds of energy in classical mechanics: kinetic energy, determined by mass and speed, and potential energy, determined by the effect of forces such as gravity. Neither of these types of energy is conserved on its own. If you drop a ball, it speeds up, thereby gaining kinetic energy. It also falls, losing potential energy. Newton's second law of motion implies that these two changes cancel each other out exactly, so the total energy does not change during the motion.

However, this is not the full story. If you put a book on a table and give it a push, its potential energy doesn't change provided the table is horizontal. But its speed does change: after an initial increase produced by the force with which you pushed it, the book quickly slows down and comes to rest. So its kinetic energy starts at a nonzero initial value just after the push, and then drops to zero. The total energy therefore also decreases, so energy is not conserved. Where has it gone? Why did the book stop? According to Newton's first law, the book should continue to move, unless some force opposes it. That force is friction between the book and the table. But what is friction?

Friction occurs when rough surfaces rub together. The rough surface of the book has bits that stick out slightly. These come into contact with parts of the table that also stick out slightly. The book pushes against the table, and the table, obeying Newton's third law, resists. This creates a force that opposes the motion of the book, so it slows down and loses energy. So where does the energy go? Perhaps conservation simply does not apply. Alternatively, the energy is still lurking somewhere, unnoticed. And that's what the first law of thermodynamics tells us: the missing energy appears as heat. Both book and table heat up slightly. Humans have known that friction creates heat even since some bright spark discovered how to rub two sticks together and start a fire. If you slide down a rope too fast, your hands get rope burns from the friction. There were plenty of clues. The first law of thermodynamics states that heat is a form of energy, and energy – thus extended – is conserved in thermodynamic processes.

The first law of thermodynamics places limits on what you can do with a heat engine. The amount of kinetic energy that you can get out, in the form of motion, cannot be more than the amount of energy you put in as heat. But it turned out that there is a further restriction on how efficiently a heat engine can convert heat energy into kinetic energy; not just the practical point that some of the energy always gets lost, but a theoretical limit that prevents all of the heat energy being converted to motion. Only some of it, the 'free' energy, can be so converted. The second law of thermodynamics turned this idea into a general principle, but it will take a while before we get to that. The limitation was discovered by Nicolas Léonard Sadi Carnot in 1824, in a simple model of how a steam engine works: the Carnot cycle.

To understand the Carnot cycle it is important to distinguish between heat and temperature. In everyday life, we say that something is hot if its temperature is high, and so confuse the two concepts. In classical thermodynamics, neither concept is straightforward. Temperature is a property of a fluid, but heat makes sense only as a measure of the transfer of energy between fluids, and is not an intrinsic property of the state (that is, the temperature, pressure, and volume) of the fluid. In the kinetic theory, the temperature of a fluid is the average kinetic energy of its molecules, and the amount of heat transferred between fluids is the change in the total kinetic energy of their molecules. In a sense heat is a bit like potential energy, which is defined relative to an arbitrary reference height; this introduces an arbitrary constant, so 'the' potential energy of a body is not uniquely defined. But when the body changes height, the difference in potential energies is the same whatever reference height is used, because the constant cancels out. In short, heat measures changes, but temperature measures states. The two are linked: heat transfer is possible only when the fluids concerned have different temperatures, and then it is transferred from the hotter one to the cooler one. This is often called the Zeroth law of thermodynamics because logically it precedes the first law, but historically it was recognised later.

Temperature can be measured using a thermometer, which exploits the expansion of a fluid, such as mercury, caused by increased temperature. Heat can be measured by using its relation to temperature. In a standard test fluid, such as water, every 1-degree rise in temperature of 1 gram of fluid corresponds to a fixed increase in the heat content. This amount is called the specific heat of the fluid, which in water is 1 calorie per gram per degree Celsius. Note that heat *increase* is a change, not a state, as required by the definition of heat.

We can visualise the Carnot cycle by thinking of a chamber containing gas, with a movable piston at one end. The cycle has four steps:

1 Heat the gas so rapidly that its temperature doesn't change. It expands, performing work on the piston.
2 Allow the gas to expand further, reducing the pressure. The gas cools.
3 Compress the gas so rapidly that its temperature doesn't change. The piston now performs work on the gas.
4 Allow the gas to expand further, increasing the pressure. The gas returns to its original temperature.

In a Carnot cycle, the heat introduced in the first step transfers kinetic energy to the piston, allowing the piston to do work. The quantity of energy transferred can be calculated in terms of the amount of heat introduced and the temperature difference between the gas and its surroundings. Carnot's theorem proves that in principle a Carnot cycle is the most efficient way to convert heat into work. This places a stringent limit on the efficiency of any heat engine, and in particular on a steam engine.

In a diagram showing the pressure and volume of the gas, a Carnot cycle looks like Figure 47 (*left*). The German physicist and mathematician Rudolf Clausius discovered a simpler way to visualise the cycle, Figure 47 (*right*). Now the two axes are temperature and a new and fundamental quantity called *entropy*. In these coordinates, the cycle becomes a rectangle, and the amount of work performed is just the area of the rectangle.

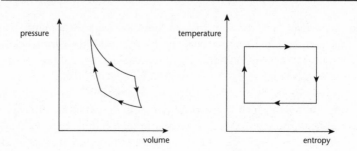

Fig 47 Carnot cycle. *Left*: In terms of pressure and volume. *Right*: In terms of temperature and entropy.

Entropy is like heat: it is defined in terms of a change of state, not a state as such. Suppose that a fluid in some initial state changes to a new state. Then the difference in entropy between the two states is the total change in the quantity 'heat divided by temperature'. In symbols, for a small step along the path between the two states, entropy S is related to heat q and temperature T by the differential equation $dS = dq/T$. The change in entropy is the change in heat per unit temperature. A large change of state can be represented as a series of small ones, so we add up all these small changes in entropy to get the overall change of entropy. Calculus tells us that the way to do this is to use an integral.[1]

Having defined entropy, the second law of thermodynamics is very simple. It states that in any physically feasible thermodynamic process, the entropy of an isolated system must always increase.[2] In symbols, $dS \geqslant 0$. For example, suppose we divide a room with a movable partition, put oxygen on one side of the partition and nitrogen on the other. Each gas has a particular entropy, relative to some initial reference state. Now remove the partition, allowing the gases to mix. The combined system also has a particular entropy, relative to the same initial reference states. And the entropy of the combined system is always greater than the sum of the entropies of the two separate gases.

Classical thermodynamics is phenomenological: it describes what you can measure, but it's not based on any coherent theory of the processes involved. That step came next with the kinetic theory of gases, pioneered by Daniel Bernoulli in 1738. This theory provides a physical explanation of pressure, temperature, the gas laws, and that mysterious quantity entropy. The basic idea – highly controversial at the time – is that a gas consists of a large number of identical molecules, which bounce around in space and occasionally collide with each other. Being a gas means that the molecules are not too tightly packed, so any given molecule spends a lot of its time travelling through the vacuum of space at a constant speed in a straight line. (I say 'vacuum' even though we're discussing a gas, because that's what the space between molecules consists of.) Since molecules, though tiny, have nonzero size, occasionally two of them will collide. Kinetic theory makes the simplifying assumption that they bounce like two colliding billiard balls, and that these balls are perfectly elastic, so no energy is lost in the collision. Among other things, this implies that the molecules keep bouncing forever.

When Bernoulli first suggested the model, the law of conservation of energy was not established and perfect elasticity seemed unlikely. The theory gradually won support from a small number of scientists, who developed their own versions and added various new ideas, but their work was almost universally ignored. The German chemist and physicist August Krönig wrote a book on the topic in 1856, simplifying the physics by not allowing the molecules to rotate. Clausius removed this simplification a year later. He claimed he had arrived at his results independently, and is now ranked as one of the first significant founders of kinetic theory. He proposed one of the key concepts of the theory, the mean free path of a molecule: how far it travels, on average, between successive collisions.

Both König and Clausius deduced the ideal gas law from kinetic theory. The three key variables are volume, pressure, and temperature. Volume is determined by the vessel that contains the gas, it sets 'boundary conditions' that affect how the gas behaves, but is not a feature of the gas as such. Pressure is the average force (per square unit of area) exerted by the molecules of the gas when they collide with the walls of the vessel. This depends on how many molecules are inside the vessel, and how fast they are moving. (They don't all move at the same speed.) Most interesting is temperature. This also depends on how fast the gas molecules are moving, and it is proportional to the average kinetic energy of the molecules. Deducing Boyle's law, the special case of the ideal gas law for constant temperature, is especially straightforward. At a fixed temperature, the distribution of velocities doesn't change, so pressure is determined by how many molecules hit the wall. If you reduce the volume, the number of molecules per cubic unit of space goes up, and the chance of any molecule hitting the wall goes up as well. Smaller volume means denser gas means more molecules hitting the wall, and this argument can be made quantitative. Similar but more complicated arguments produce the ideal gas law in all its glory as long as the molecules aren't squashed too tightly together. So now there was a deeper theoretical basis for Boyle's law, based on the theory of molecules.

Maxwell was inspired by Clausius's work, and in 1859 he placed kinetic theory on mathematical foundations by writing down a formula for the probability that a molecule will travel with a given speed. It is based on the normal distribution or bell curve (Chapter 7). Maxwell's formula seems to have been the first instance of a physical law based on probabilities. He was followed by the Austrian physicist Ludwig Boltzmann, who developed the same formula, now called the Maxwell–Boltzmann distribution. Boltzmann reinterpreted thermodynamics in terms of the kinetic theory of gases, founding what is now called statistical mechanics. In particular, he came up with a new interpretation of entropy, relating the thermodynamic concept to a statistical feature of the molecules in the gas.

The traditional thermodynamic quantities, such as temperature, pressure, heat, and entropy, all refer to large-scale average properties of the gas. However, the fine structure consists of lots of molecules whizzing around and bumping into each other. The same large-scale state can arise from innumerable different small-scale states, because minor differences on the small scale average out. Boltzmann therefore distinguished macrostates and microstates of the system: large-scale averages and the actual states of the molecules. Using this, he showed that entropy, a macrostate, can be interpreted as a statistical feature of microstates. He expressed this in the equation

$$S = k \log W$$

Here S is the entropy of the system, W is the number of distinct microstates that can give rise to the overall macrostate, and k is a constant. It is now called Boltzmann's constant, and its value is 1.38×10^{-23} joules per degree kelvin.

It is this formula that motivates the interpretation of entropy as disorder. The idea is that fewer microstates correspond to an ordered macrostate than to a disordered one, and we can understand why by thinking about a pack of cards. For simplicity, suppose that we have just six cards, marked 2, 3, 4, J, Q, K. Put them in two separate piles, with the low-value cards in one pile and the court cards in the other. This is an ordered arrangement. In fact, it retains traces of order if you shuffle each pile, but keep the piles separate, because however you do this, the low-value cards are all in one pile and the court cards are in the other. However, if you shuffle both piles together, the two types of card can become mixed, with arrangements like 4QK2J3. Intuitively, these mixed-up arrangements are more disordered.

Let's see how this relates to Boltzmann's formula. There are 36 ways to arrange the cards in their two piles: six for each pile. But there are 720 ways ($6! = 1 \times 2 \times 3 \times 4 \times 5 \times 6$) to arrange all six cards in order. The type of ordering of the cards that we allow – two piles or one – is analogous to the macrostate of a thermodynamic system. The exact order is the microstate. The more ordered macrostate has 36 microstates, the less ordered one has 720. So the more microstates there are, the less ordered the corresponding macrostate becomes. Since logarithms get bigger when the numbers do, the greater the logarithm of the number of microstates, the more disordered the macrostate becomes. Here

$$\log 36 = 3.58 \qquad \log 720 = 6.58$$

These are effectively the entropies of the two macrostates. Boltzmann's constant just scales the values to fit the thermodynamic formalism when we're dealing with gases.

The two piles of cards are like two non-interacting thermodynamic states, such as a box with a partition separating two gases. Their individual entropies are each log 6, so the total entropy is 2 log 6, which equals log 36. So the logarithm makes entropy *additive* for non-interacting systems: to get the entropy of the combined (but not yet interacting) system, add the separate entropies. If we now let the systems interact (remove the partition) the entropy increases to log 720.

The more cards there are, the more pronounced this effect becomes. Split a standard pack of 52 playing cards into two piles, with all the red cards in one pile and all the black cards in the other. This arrangement can occur in $(26!)^2$ ways, which is about 1.62×10^{53}. Shuffling both piles together we get 52! microstates, roughly 8.07×10^{67}. The logarithms are 122.52 and 156.36 respectively, and again the second is larger.

Boltzmann's ideas were not received with great acclaim. At a technical level, thermodynamics was beset with difficult conceptual issues. One was the precise meaning of 'microstate'. The position and velocity of a molecule are continuous variables, able to take on infinitely many values, but Boltzmann needed a finite number of microstates in order to count how many there were and then take the logarithm. So these variables had to be 'coarse-grained' in some manner, by splitting the continuum of possible values into finitely many very small intervals.

Another issue, more philosophical in nature, was the arrow of time – an apparent conflict between the time-reversible dynamics of microstates and the one-way time of macrostates, determined by entropy increase. The two issues are related, as we will shortly see.

The biggest obstacle to the theory's acceptance, however, was the idea that matter is made from extremely tiny particles, atoms. This concept, and the word atom, which means 'indivisible', goes back to ancient Greece, but even around 1900 the majority of physicists did not believe that matter is made from atoms. So they didn't believe in molecules, either, and a theory of gases based on them was obviously nonsense. Maxwell, Boltzmann, and other pioneers of kinetic theory were convinced that molecules and atoms were real, but to the skeptics, atomic theory was just a convenient way to picture matter. No atoms had ever been observed, so there was no scientific evidence that they existed. Molecules, specific combinations of atoms, were similarly controversial. Yes, atomic theory fitted all sorts of experimental data in chemistry, but that was not proof that atoms existed.

One of the things that finally convinced most objectors was the use of kinetic theory to make predictions about Brownian motion. This effect was discovered by a Scottish botanist, Robert Brown.[3] He pioneered the use of the microscope, discovering, among other things, the existence of the nucleus of a cell, now known to be the repository of its genetic information. In 1827 Brown was looking through his microscope at pollen grains in a fluid, and he spotted even tinier particles that had been ejected by the pollen. These tiny particles jiggled around in a random manner, and at first Brown wondered if they were some diminutive form of life. However, his experiments showed the same effect in particles derived from non-living matter, so whatever caused the jiggling, it didn't have to be alive. At the time, no one knew what caused this effect. We now know that the particles ejected by the pollen were organelles, tiny subsystems of the cell with specific functions; in this case, to manufacture starch and fats. And we interpret their random jiggles as evidence for the theory that matter is made from atoms.

The link to atoms comes from mathematical models of Brownian motion, which first turned up in statistical work of the Danish astronomer and actuary Thorvald Thiele in 1880. The big advance was made by Einstein in 1905 and the Polish scientist Marian Smoluchowski in 1906. They independently proposed a physical explanation of Brownian motion: atoms of the fluid in which the particles were floating were randomly bumping into the particles and giving them tiny kicks. On this basis, Einstein used a mathematical model to make quantitative predictions about the statistics of the motion, which were confirmed by Jean Baptiste Perrin in 1908–9.

Boltzmann committed suicide in 1906 – just when the scientific world was starting to appreciate that the basis of his theory was real.

In Boltzmann's formulation of thermodynamics, molecules in a gas are analogous to cards in a pack, and the natural dynamics of the molecules is analogous to shuffling. Suppose that at some moment all the oxygen molecules in a room are concentrated at one end, and all the nitrogen molecules are at the other. This is an ordered thermodynamic state, like two separate piles of cards. After a very short period, however, random collisions will mix all the molecules together, more or less uniformly throughout the room, like shuffling the cards. We've just seen that this process typically causes entropy to increase. This is the orthodox picture of the relentless increase of entropy, and it is the standard interpretation of the second law: 'the amount of disorder in the universe steadily increases'. I'm pretty sure that this characterisation of the second law would have satisfied Snow if anyone had offered it. In this form, one dramatic consequence of the second law is the scenario of the 'heat death of the universe', in which the entire universe will eventually become a lukewarm gas with no interesting structure whatsoever.

Entropy, and the mathematical formalism that goes with it, provides an excellent model for many things. It explains why heat engines can only reach a particular level of efficiency, which prevents engineers wasting valuable time and money looking for a mare's nest. That's not just true of Victorian steam engines, it applies to modern car engines as well. Engine design is one of the practical areas that has benefited from knowing the laws of thermodynamics. Refrigerators are another. They use chemical reactions to transfer heat out of the food in the fridge. It has to go somewhere: you can often feel the heat rising from the outside of the fridge's motor housing. The same goes for air-conditioning. Power generation is another application. In a coal, gas, or nuclear power station, what it initially generated is heat. The heat creates steam, which drives a turbine. The turbine, following principles that go back to Faraday, turns motion into electricity.

The second law of thermodynamics also governs the amount of energy we can hope to extract from renewable resources, such as wind and waves. Climate change has added new urgency to this question, because renewable energy sources produce less carbon dioxide than conventional ones. Even nuclear power has a big carbon footprint, because the fuel has to be made, transported, and stored when it is no longer useful but still radioactive. As I write there is a simmering debate about the maximum amount of energy that we can extract from the ocean and the atmosphere without causing the kinds of change that we are hoping to avoid. It is based on thermodynamic estimates of the amount of free energy in those natural systems. This is an important issue: if renewables *in principle* cannot supply the energy we need, then we have to look elsewhere. Solar panels, which extract energy directly from sunlight, are not directly affected by the thermodynamic limits, but even those involve manufacturing processes and so on. At the moment, the case that such limits are a serious obstacle relies on some sweeping simplifications, and even if they are correct, the calculations do not rule out renewables as a source for most of the world's power. But it's worth remembering that similarly broad calculations about carbon dioxide production, performed in the 1950s, have proved surprisingly accurate as a predictor of global warming.

The second law works brilliantly in its original context, the behaviour of gases, but it seems to conflict with the rich complexities of our planet, in particular, life. It seems to rule out the complexity and organisation exhibited by living systems. So the second law is sometimes invoked to attack Darwinian evolution. However, the physics of steam engines is not particularly appropriate to the study of life. In the kinetic theory of gases, the forces that act between the molecules are short-range (active only when the molecules collide) and repulsive (they bounce). But most of the forces of nature aren't like that. For example, gravity acts at enormous distances, and it is attractive. The expansion of the universe away from the Big Bang has not smeared matter out into a uniform gas. Instead, the matter has formed into clumps – planets, stars, galaxies, supergalactic clusters... The forces that hold molecules together are also attractive – except at very short distances where they become repulsive, which stops the molecule collapsing – but their effective range is fairly short. For systems such as these, the thermodynamic model of independent subsystems whose interactions switch on but not off is simply irrelevant. The features of thermodynamics either don't apply, or are so long-term that they don't model anything interesting.

The laws of thermodynamics, then, underlie many things that we take for granted. And the interpretation of entropy as 'disorder' helps us to understand those laws and gain an intuitive feeling for their physical basis.

However, there are occasions when interpreting entropy as disorder seems to lead to paradoxes. This is a more philosophical realm of discourse – and it's fascinating.

One of the deep mysteries of physics is time's arrow. Time seems to flow in one particular direction. However, it seems logically and mathematically possible for time to flow backwards instead – a possibility exploited by books such as Martin Amis's *Time's Arrow*, the much earlier novel *Counter-Clock World* by Philip K. Dick, and the BBC television series *Red Dwarf*, whose protagonists memorably drank beer and engaged in a bar brawl in reverse time. So why can't time flow the other way? At first sight, thermodynamics offers a simple explanation for the arrow of time: it is the direction of entropy increase. Thermodynamic processes are irreversible: oxygen and nitrogen will spontaneously mix, but not spontaneously unmix.

There is a puzzle here, however, because any classical mechanical system, such as the molecules in a room, is time-reversible. If you keep shuffling a pack of cards at random, then eventually it will get back to its original order. In the mathematical equations, if at some instant the velocities of all particles are simultaneously reversed, then the system will retrace its steps, back-to-front in time. The entire universe can bounce, obeying the same equations in both directions. So why do we never see an egg unscrambling?

The usual thermodynamic answer is: a scrambled egg is more disordered than an unscrambled one, entropy increases, and that's the way time flows. But there's a subtler reason why eggs don't unscramble: the universe is very, very unlikely to bounce in the required manner. The probability of that happening is ridiculously small. So the discrepancy between entropy increase and time–reversibility comes from the initial conditions, not the equations. The equations for moving molecules are time-reversible, but the initial conditions are not. When we reverse time, we must use 'initial' conditions given by the *final* state of the forward-time motion.

The most important distinction here is between symmetry of equations and symmetry of their solutions. The equations for bouncing molecules have time-reversal symmetry, but individual solutions can have a definite arrow of time. The most you can deduce about a solution, from time-reversibility of the equation, is that there must also exist *another* solution that is the time-reversal of the first. If Alice throws a ball to Bob, the time-reversed solution has Bob throwing a ball to Alice. Similarly, since the equations of mechanics allow a vase to fall to the ground and smash into a thousand pieces, they must also allow a solution in which a thousand shards of glass mysteriously move together, assemble themselves into an intact vase, and leap into the air.

There's clearly something funny going on here, and it repays investigation. We don't have a problem with Bob and Alice tossing a ball either way. We see such things every day. But we don't see a smashed vase putting itself back together. We don't see an egg unscrambling.

Suppose we smash a vase and film the result. We start with a simple, ordered state – an intact vase. It falls to the floor, where the impact breaks it into pieces and propels those pieces all over the floor. They slow down and come to a halt. It all looks entirely normal. Now play the movie backwards. Bits of glass, which just happen to be the right shape to fit together, are lying on the floor. Spontaneously, they start to move. They move at just the right speed, and in just the right direction, to meet. They assemble into a vase, which heads skywards. That doesn't seem right.

In fact, as described, it's not right. Several laws of mechanics appear to be violated, among them conservation of momentum and conservation of energy. Stationary masses can't suddenly move. A vase can't gain energy from nowhere and leap into the air.

Ah, yes... but that's because we're not looking carefully enough. The vase didn't leap into the air of its own accord. The floor started to vibrate, and the vibrations came together to give the vase a sharp kick into the air. The bits of glass were similarly impelled to move by incoming waves of vibration of the floor. If we trace those vibrations back, they spread out, and seem to die down. Eventually friction dissipates all movement... Oh, yes, friction. What happens to kinetic energy when there's friction? It turns into heat. So we've missed some details of the time-reversed scenario. Momentum and energy do balance, but the missing amounts come from the floor losing heat.

In principle, we could set up a forward-time system to mimic the time-reversed vase. We just have to arrange for molecules in the floor to collide in just the right way to release some of their heat as motion of the floor, kick the pieces of glass in just the right way, then hurl the vase into the air. The point is not that this is impossible in principle: if it were, time-reversibility would fail. But it's impossible in practice, because there is no way to control that many molecules that precisely.

This, too, is an issue about boundary conditions – in this case, initial conditions. The initial conditions for the vase-smashing experiment are easy to implement, and the apparatus is easy to acquire. It's all very robust, too: use another vase, drop it from a different height... much the same will happen. The vase-assembling experiment, in contrast, requires extraordinarily precise control of gazillions of individual molecules and exquisitely carefully made pieces of glass. Without all that control equipment disturbing a single molecule. That's why we can't actually do it.

However, notice how we're thinking here: we're focusing on *initial* conditions. That sets up an arrow of time: the rest of the action comes later than the start. If we looked at the vase-smashing experiment's *final* conditions, right down to the molecular level, they would be so complex that no one in their right mind would even consider trying to replicate them.

The mathematics of entropy fudges out these very small scale considerations. It allows vibrations to die away but not to increase. It allows friction to turn into heat but not heat to turn into friction. The discrepancy between the second law of thermodynamics and microscopic reversibility arises from coarse-graining, the modelling assumptions made when passing from a detailed molecular description to a statistical one. These assumptions implicitly specify an arrow of time: large-scale disturbances are allowed to die down below the perceptible level *as time passes*, but small-scale disturbances are not allowed to follow the time-reversed scenario. Once the dynamics passes through this temporal trapdoor, it's not allowed to come back.

If entropy always increases, how did the chicken ever create the ordered egg to begin with? A common explanation, advanced by the Austrian physicist Erwin Schrödinger in 1944 in a brief and charming book *What is Life?*, is that living systems somehow borrow order from their environment, and pay it back by making the environment even more disordered than it would otherwise have been. This extra order corresponds to 'negative entropy', which the chicken can use to make an egg without violating the second law. In Chapter 15 we will see that negative entropy can, in appropriate circumstances, be thought of as information, and it is often claimed that the chicken accesses information – provided by its DNA, for example – to obtain the necessary negative entropy. However, the identification of information with negative entropy makes sense only in very specific contexts, and the activities of living creatures are not one of them. Organisms create order through the processes that they carry out, but those processes are not thermodynamic. Chickens don't access some storehouse of order to make the thermodynamic books balance: they use processes for which a thermodynamic model is inappropriate, and throw the books away because they don't apply.

The scenario in which an egg is created by borrowing entropy would be appropriate if the process that the chicken used were the time-reversal of an egg breaking up into its constituent molecules. At first sight this is vaguely plausible, because the molecules that eventually form the egg are scattered throughout the environment; they come together in the chicken, where biochemical processes put them together in an ordered manner to form the egg. However, there is a difference in the initial conditions. If you went round beforehand labelling molecules in the chicken's environment, to say 'this one will end up in the egg at such and such a location', you would in effect be creating initial conditions as complex and unlikely as those for unscrambling an egg. But that's not how the chicken operates. Some molecules happen to end up in the egg and are conceptually labelled as part of it *after* the process is complete. Other molecules could have done the same job – one molecule of calcium carbonate is just as good for making a shell as any other. So the chicken is not creating order from disorder. The order is assigned to the end result of the egg-making process – like shuffling a pack of cards into a random order and then numbering them 1, 2, 3, and so on with a felt-tipped pen. Amazing – they're in numerical order!

To be sure, the egg looks more ordered than its ingredients, even if we take account of this difference in initial conditions. But that's because the process that makes an egg is not thermodynamic. Many physical processes do, in effect, unscramble eggs. An example is the way minerals dissolved in water can create stalactites and stalagmites in caves. If we specified the exact form of stalactite we wanted, ahead of time, we'd be in the same position as someone trying to unsmash a vase. But if we're willing to settle for any old stalactite, we get one: order from disorder. Those two terms are often used in a sloppy way. What matters are what kind of order and what kind of disorder. That said, I *still* don't expect to see an egg unscrambling. There is no feasible way to set up the necessary initial conditions. The best we can do is turn the scrambled egg into chickenfeed and wait for the bird to lay a new one.

In fact, there is a reason why we wouldn't see an egg unscrambling, even if the world did run backwards. Because we and our memories are part of the system that is being reversed, we wouldn't be sure which way time was 'really' running. Our sense of the flow of time is produced by memories, physico-chemical patterns in the brain. In conventional language, the brain stores records of the past but not of the future.

Imagine making a series of snapshots of the brain watching an egg being scrambled, along with its memories of the process. At one stage the brain remembers a cold, unscrambled egg, and some of its history when taken from the fridge and put into the saucepan. At another stage it remembers having whisked the egg with a fork, and having moved it from the fridge to the saucepan.

If we now run the entire universe in reverse, we reverse the order in which those memories occur, in 'real' time. But we don't reverse the ordering of a given memory in the brain. At the start (in reversed time) of the process that unscrambles the egg, the brain does not remember the 'past' of that egg – how it emerged from a mouth on to a spoon, was unwhisked, gradually building up a complete egg ... Instead, the record in the brain at that moment is one in which it remembers having cracked open an egg, along with the process of moving it from the fridge to the saucepan and scrambling it. But this memory is exactly the same as one of the records in the forward-time scenario. The same goes for all the other memory snapshots. Our perception of the world depends on what we observe *now*, and what memories our brain holds, *now*. In a time-reversed universe, we would in effect remember the future, not the past.

The paradoxes of time-reversibility and entropy are not problems about the real world. They are problems about the assumptions we make when we try to model it.

13 One thing is absolute
Relativity

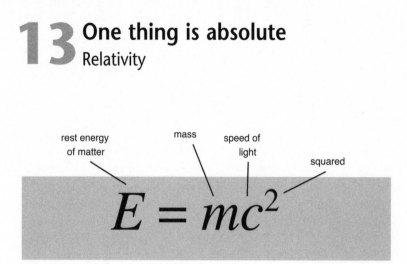

rest energy of matter   mass   speed of light   squared

$$E = mc^2$$

What does it say?

Matter contains energy equal to its mass multiplied by the square of the speed of light.

Why is that important?

The speed of light is huge and its square is absolutely humongous. One kilogram of matter would release about 40% of the energy in the largest nuclear weapon ever exploded. It's part of a package of equations that changed our view of space, time, matter, and gravity.

What did it lead to?

Radical new physics, definitely. Nuclear weapons... well, just maybe – though not as directly or conclusively as the urban myths claim. Black holes, the Big Bang, GPS and satnav.

J ust as Albert Einstein, with his startled mop hairdo, is the archetypal scientist in popular culture, so his equation $E = mc^2$ is the archetypal equation. It is widely believed that the equation led to the invention of nuclear weapons, that it comes from Einstein's theory of relativity, and that this theory (obviously) has something to do with various things being relative. In fact, many social relativists happily chant 'everything is relative', and think it has something to do with Einstein.

It doesn't. Einstein named his theory 'relativity' because it was a modification of the rules for relative motion that had traditionally been used in Newtonian mechanics, where motion *is* relative, depending in a very simple and intuitive way on the frame of reference in which it is observed. Einstein had to tweak Newtonian relativity to make sense of a baffling experimental discovery: that one particular physical phenomenon is not relative at all, but absolute. From this he derived a new kind of physics in which objects shrink when they move very fast, time slows to a crawl, and mass increases without limit. An extension incorporating gravity has given us the best understanding we yet have of the origins of the universe and the structure of the cosmos. It is based on the idea that space and time can be curved.

Relativity is real. The Global Positioning System (GPS, used among other things for car satnav) works only when corrections are made for relativistic effects. The same goes for particle accelerators such as the Large Hadron Collider, currently searching for the Higgs boson, thought to be the origin of mass. Modern communications have become so fast that market traders are beginning to run up against a relativistic limitation: the speed of light. This is the fastest that any message, such as an Internet instruction to buy or sell stock, can travel. Some see this as an opportunity to cut a deal nanoseconds earlier than the competition, but so far, relativistic effects haven't had a serious effect on international finance. However, people have already worked out the best locations for new stock markets or dealerships. It's only a matter of time.

At any rate, not only is relativity not relative: even the iconic equation is not what it seems. When Einstein first derived the physical idea that it represents, he didn't write it in the familiar way. It is not a mathematical consequence of relativity, though it becomes one if various physical assumptions and definitions are accepted. It is perhaps typical of human culture that our most iconic equation is not, and was not, what it seems to be, and neither is the theory that gave birth to it. Even the connection with nuclear weapons is not clear-cut, and its historical influence on the first atomic bomb was small compared with Einstein's political clout as *the* iconic scientist.

'Relativity' covers two distinct but related theories: special relativity and general relativity. I'll use Einstein's celebrated equation as an excuse to talk about both. Special relativity is about space, time, and matter in the absence of gravity; general relativity takes gravity into account as well. The two theories are part of one big picture, but it took Einstein ten years of intensive effort to discover how to modify special relativity to incorporate gravity. Both theories were inspired by difficulties in reconciling Newtonian physics with observations, but the iconic formula arose in special relativity.

Physics seemed fairly straightforward and intuitive in Newton's day. Space was space, time was time, and never the twain should meet. The geometry of space was that of Euclid. Time was independent of space, the same for all observers – provided they had synchronised their clocks. The mass and size of a body did not change when it moved, and time always passed at the same rate everywhere. But when Einstein had finished reformulating physics, all of these statements – so intuitive that it is very difficult to imagine how any of them could fail to represent reality – turned out to be wrong.

They were not totally wrong, of course. If they had been nonsense, then Newton's work would never have got off the ground. The Newtonian picture of the physical universe is an approximation, not an exact description. The approximation is extremely accurate provided everything involved is moving slowly enough, and in most everyday circumstances that is the case. Even a jet fighter, travelling at twice the speed of sound, is moving slowly for this purpose. But one thing that does play a role in everyday life moves very fast indeed, and sets the yardstick for all other speeds: light. Newton and his successors had demonstrated that light was a wave, and Maxwell's equations confirmed this. But the wave nature of light raised a new issue. Ocean waves are waves in water, sound waves are waves in air, earthquakes are waves in the Earth. So light waves are waves in... what?

Mathematically they are waves in the electromagnetic field, which is assumed to pervade the whole of space. When the electromagnetic field is excited – persuaded to support electricity and magnetism – we observe a wave. But what happens when it's *not* excited? Without waves, an ocean would still be an ocean, air would still be air, and the Earth would still be the Earth. Analogously, the electromagnetic field would still be... the electromagnetic field. But you can't observe the electromagnetic field if there's no electricity or magnetism going on. If you can't observe it, what is it? Does it exist at all?

All waves known to physics, except the electromagnetic field, are waves in something tangible. All three types of wave – water, air, earthquake – are waves of movement. The medium moves up and down or from side to side, but usually it doesn't travel with the wave. (Tie a long rope to a wall and shake one end: a wave travels along the rope. But the *rope* doesn't travel along the rope.) There are exceptions: when air travels along with the wave we call it 'wind', and ocean waves move water up a beach when they hit one. But even though we describe a tsunami as a moving wall of water, it doesn't roll across the top of the ocean like a football rolling along the pitch. Mostly, the water in any given location goes up and down. It is the location of the 'up' that moves. Until the water gets close to shore; then you get something much more like a moving wall.

Light, and electromagnetic waves in general, didn't seem to be waves in anything tangible. In Maxwell's day, and for fifty years or more afterwards, that was disturbing. Newton's law of gravity had long been criticised because it implies that gravity somehow 'acts at a distance', as miraculous in philosophical principle as kicking a ball into the goal when you're sitting in the stands. Saying that it is transmitted by 'the gravitational field' doesn't really explain what's happening. The same goes for electromagnetism. So physicists came round to the idea that there was some medium – no one knew what, they called it the 'luminiferous aether' or just plain 'ether' – that supported electromagnetic waves. Vibrations travel faster the more rigid the medium, and light was very fast indeed, so the ether had to be extremely rigid. Yet planets could move through it without resistance. To have avoided easy detection, the ether must have no mass, no viscosity, be incompressible, and be totally transparent to all forms of radiation.

It was a daunting combination of attributes, but almost all physicists assumed the ether existed, because light clearly did what light did.

Something had to carry the wave. Moreover, the ether's existence could in principle be detected, because another feature of light suggested a way to observe it. In a vacuum, light moves with a fixed speed c. Newtonian mechanics had taught every physicist to ask: speed relative to what? If you measure a velocity in two different frames of reference, one moving with respect to the other, you get different answers. The constancy of the speed of light suggested an obvious reply: *relative to the ether*. But this was a little facile, because two frames of reference that are moving with respect to each other can't both be at rest relative to the ether.

As the Earth ploughs its way through the ether, miraculously unresisted, it goes round and round the Sun. At opposite points on its orbit it is moving in opposite directions. So by Newtonian mechanics, the speed of light should vary between two extremes: c plus a contribution from the Earth's motion relative to the ether, and c minus the same contribution. Measure the speed, measure it six months later, find the difference: if there is one, you have proof that the ether exists. In the late 1800s many experiments along these lines were carried out, but the results were inconclusive. Either there was no difference, or there was one but the experimental method wasn't accurate enough. Worse, the Earth might be dragging the ether along with it. This would simultaneously explain why the Earth can move through such a rigid medium without resistance, and imply that you ought not to see any difference in the speed of light anyway. The Earth's motion relative to the ether would always be zero.

In 1887 Albert Michelson and Edward Morley carried out one of the most famous physics experiments of all time. Their apparatus was designed to detect extremely small variations in the speed of light in two directions, at right angles to each other. However the Earth was moving relative to the ether, it couldn't be moving with the same relative speed in two different directions... unless it happened by coincidence to be moving along the line bisecting those directions, in which case you just rotated the apparatus a little and tried again.

The apparatus, Figure 48, was small enough to fit on a laboratory desk. It used a half-silvered mirror to split a beam of light into two parts, one passing through the mirror and the other being reflected at right angles. Each separate beam was reflected back along its path, and the two beams combined again, to hit a detector. The apparatus was adjusted to make the paths the same length. The original beam was set up to be coherent, meaning that its waves were in synchrony with each other – all having the same phase, peaks coinciding with peaks. Any difference between the speed of light in the directions followed by the two beams would cause

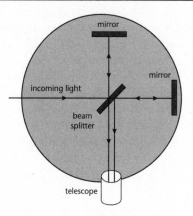

Fig 48 The Michelson–Morley experiment.

their phases to shift relative to each other, so their peaks would be in different places. This would cause interference between the two waves, resulting in a striped pattern of 'diffraction fringes'. Motion of the Earth relative to the ether would cause the fringes to move. The effect would be tiny: given what was known about the Earth's motion relative to the Sun, the diffraction fringes would shift by about 4% of the width of one fringe. By using multiple reflections, this could be increased to 40%, big enough to be detected. To avoid the possible coincidence of the Earth moving exactly along the bisector of the two beams, Michelson and Morley floated the apparatus on a bath of mercury, so that it could be spun round easily and rapidly. It should then be possible to watch the fringes shifting with equal rapidity.

It was a careful, accurate experiment. Its result was entirely negative. The fringes did not move by 40% of their width. As far as anyone could tell with certainty, they didn't move at all. Later experiments, capable of detecting a shift 0.07% as wide as a fringe, also gave a negative result. The ether did not exist.

This result didn't just dispose of the ether: it threatened to dispose of Maxwell's theory of electromagnetism, too. It implied that light does not behave in a Newtonian manner, relative to moving frames of reference. This problem can be traced right back to the mathematical properties of Maxwell's equations and how they transform relative to a moving frame. The Irish physicist and chemist George FitzGerald and the Dutch physicist Hendrik Lorenz independently suggested (in 1892 and 1895 respectively) an audacious way to get round the problem. If a moving body contracts slightly in its direction of motion, by just the right amount, then the change in phase that the Michelson–Morley experiment was hoping to detect would be exactly cancelled out by the change in the length of the path that the light was following. Lorenz showed that this 'Lorenz–FitzGerald contraction' sorted out the mathematical difficulties with the Maxwell equations as well. The joint discovery showed that the results of experiments on electromagnetism, including light, do not depend on the relative motion of the reference frame. Poincaré, who had also been working along similar lines, added his persuasive intellectual weight to the idea.

The stage was now set for Einstein. In 1905 he developed and extended previous speculations about a new theory of relative motion in a paper 'On the electrodynamics of moving bodies'. His work went beyond that of his predecessors in two ways. He showed that the necessary change to the mathematical formulation of relative motion was more than just a trick to sort out electromagnetism. It was required for all physical laws. It followed that the new mathematics must be a genuine description of reality, with the same philosophical status that had been accorded to the prevailing Newtonian description, but providing a better agreement with experiments. It was real physics.

The view of relative motion employed by Newton went back even further, to Galileo. In his 1632 *Dialogo sopra i due massimi sistemi del mondo* ('Dialogue Concerning the Two Chief World Systems') Galileo discussed a ship travelling at constant velocity on a perfectly smooth sea, arguing that no experiment in mechanics carried out below decks could reveal that the ship was moving. This is Galileo's principle of relativity: in mechanics, there is no difference between observations made in two frames that are moving with uniform velocity relative to each other. In particular, there is no special frame of reference that is 'at rest'. Einstein's starting-point was the same principle, but with an extra twist: it must apply not just to mechanics, but to all physical laws. Among them, of course, being Maxwell's equations and the constancy of the speed of light.

To Einstein, the Michelson–Morley experiment was a small piece of extra evidence, but it wasn't proof of the pudding. The proof that his new theory was correct lay in his extended principle of relativity, and what it implied for the mathematical structure of the laws of physics. If you accepted the principle, all else followed. This is why the theory became known as 'relativity'. Not because 'everything is relative', but because you have to take into account the *manner* in which everything is relative. And it's not what you expect.

This version of Einstein's theory is known as special relativity because it applies only to frames of reference that are moving uniformly with respect to each other. Among its consequences are the Lorenz–FitzGerald contraction, now interpreted as a necessary feature of space-time. In fact, there were three related effects. If one frame of reference is moving uniformly relative to another one, then lengths measured in that frame contract along the direction of motion, masses increase, and time runs more slowly. These three effects are tied together by the basic conservation laws of energy and momentum; once you accept one of them, the others are logical consequences.

The technical formulation of these effects is a formula that describes how measurements in one frame relate to those in the other. The executive summary is: if a body could move close to the speed of light, then its length would become very small, time would slow to a crawl, and its mass would become very large. I'll just give a flavour of the mathematics: the physical description should not be taken too literally and it would take too long to set it up in the correct language. It all comes from... Pythagoras's theorem. One of the oldest equations in science leads to one of the newest.

Suppose that a spaceship is passing overhead with velocity v, and the crew performs an experiment. They send a pulse of light from the floor of the cabin to the roof, and measure the time taken to be T. Meanwhile an observer on the ground watches the experiment through a telescope (assume the spaceship is transparent), measuring the time to be t.

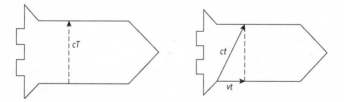

Fig 49 *Left:* The experiment in the crew's frame of reference. *Right:* The same experiment in the ground observer's frame of reference. Grey shows the ship's position as seen from the ground when the light beam starts its journey; black shows the ship's position when the light completes its journey.

Figure 49 *(left)* shows the geometry of the experiment from the crew's point of view. To them, the light has gone straight up. Since light travels at speed c, the distance travelled is cT, shown by the dotted arrow. Figure 49 *(right)* shows the geometry of the experiment from the ground observer's point of view. The spaceship has moved a distance vt, so the light has travelled diagonally. Since light *also* travels at speed c for the ground observer, the diagonal has length ct. But the dotted line has the same length as the dotted arrow in the first picture, namely cT. By Pythagoras's theorem,

$$(ct)^2 = (CT)^2 + (vt)^2$$

We solve for T, getting

$$T = t\sqrt{1 - \frac{v^2}{c^2}}$$

which is smaller than t.

To derive the Lorenz–FitzGerald contraction, we now imagine that the spaceship travels to a planet distance x from Earth at speed v. Then the elapsed time is $t = x/v$. But the previous formula shows that to the crew, the time taken is T, not t. For them, the distance X must satisfy $T = X/v$. Therefore

$$X = x\sqrt{1 - \frac{v^2}{c^2}}$$

which is smaller than x.

The derivation of the mass change is slightly more involved, and it depends on a particular interpretation of mass, 'rest mass', so I won't give details. The formula is

$$M = m \Big/ \sqrt{1 - \frac{v^2}{c^2}}$$

which is larger than m.

These equations tell us that there is something very special about the speed of light (and indeed about light). An important consequence of this formalism is that the speed of light is an impenetrable barrier. If a body starts out slower than light, it cannot be accelerated to a speed greater than that of light. In September 2011 physicists working in Italy announced that subatomic particles called neutrinos appeared to be travelling faster than light.[1] Their observation is controversial, but if it is confirmed, it will lead to important new physics.

Pythagoras turns up in relativity in other ways. One is the formulation of special relativity in terms of the geometry of space-time, originally introduced by Hermann Minkowski. Ordinary Newtonian space can be captured mathematically by making its points correspond to three coordinates (x, y, z), and defining the distance d between such a point and another one (X, Y, Z) using Pythagoras's theorem:

$$d^2 = (x - X)^2 + (y - Y)^2 + (z - Z)^2$$

Now take the square root to get d. Minkowski space-time is similar, but now there are four coordinates (x, y, z, t), three of space plus one of time, and a point is called an *event* – a location in space, observed at a specified time. The distance formula is very similar:

$$d^2 = (x - X)^2 + (y - Y)^2 + (z - Z)^2 - c^2(t - T)^2$$

The factor c^2 is just a consequence of the units used to measure time, but the minus sign in front of it is crucial. The 'distance' d is called the interval, and the square root is real only when the right-hand side of the equation is positive. That boils down to the spatial distance between the two events being smaller than the temporal difference (in correct units: light-years and years, for instance). That in turn means that in principle a body could travel from the first point in space at the first time, and arrive at the second point in space at the second time, without going faster than light.

In other words, the interval is real if and only if it is physically possible, in principle, to travel between the two events. The interval is zero if and only if light could travel between them. This physically accessible region is called the light cone of an event, and it comes in two pieces: the past and the future. Figure 50 shows the geometry when space is reduced to one dimension.

I've now shown you three relativistic equations, and sketched how they arose, but none of them is Einstein's iconic equation. However, we're now ready to understand how he derived it, once we appreciate one more innovation of early twentieth-century physics. As we've seen, physicists had previously performed experiments to demonstrate conclusively that light is a wave, and Maxwell had shown that it is a wave of electromagnetism. However, by 1905 it was becoming clear that despite the weight of evidence for the wave nature of light, there are circumstances in which it behaves like a particle. In that year Einstein used this idea to explain some features of the photoelectric effect, in which light that hits a

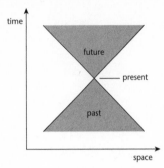

Fig 50 Minkowski space-time, with space shown as one-dimensional.

suitable metal generates electricity. He argued that the experiments made sense only if light comes in discrete packages: in effect, particles. They are now called photons.

This puzzling discovery was one of the key steps towards quantum mechanics, and I'll say more about it in Chapter 14. Curiously, this quintessentially quantum-mechanical idea was vital to Einstein's formulation of relativity. To derive his equation relating mass to energy, Einstein thought about what happens to a body that emits a pair of photons. To simplify the calculations he restricted attention to one dimension of space, so that the body moved along a straight line. This simplification does not affect the answer. The basic idea is to consider the system in two different frames of reference.[2] One moves with the body, so that the body appears to be stationary within that frame. The other frame moves with a small, nonzero velocity relative to the body. Let me call these the stationary and moving frames. They are like the spaceship (in its own frame of reference it is stationary) and my ground observer (to whom it appears to be moving).

Einstein assumed that the two photons are equally energetic, but emitted in opposite directions. Their velocities are equal and opposite, so the velocity of the body (in either frame) does not change when the photons are emitted. Then he calculated the energy of the system before the body emits the pair of photons, and afterwards. By assuming that energy must be conserved, he obtained an expression that relates the change in the body's energy, caused by emitting the photons, to the change in its (relativistic) mass. The upshot was:

$$(\text{change in energy}) = (\text{change in mass}) \times c^2$$

Making the reasonable assumption that a body of zero mass has zero energy, it then followed that

$$\text{energy} = \text{mass} \times c^2$$

This, of course, is the famous formula, in which E symbolises energy and m mass.

As well as doing the calculations, Einstein had to interpret their meaning. In particular, he argued that in a frame for which the body is at rest, the energy given by the formula should be considered to be its 'internal' energy, which it possesses because it is made from subatomic particles, each of which has its own energy. In a moving frame, there is also a contribution from kinetic energy. There are other mathematical subtleties too, such as the use of a small velocity and approximations to the exact formulas.

Einstein is often credited, if that's the word, with the realisation that an atomic bomb would release stupendous quantities of energy. Certainly *Time* magazine gave that impression in July 1946 when it put his face on the cover with an atomic mushroom cloud in the background bearing his iconic equation. The connection between the equation and a huge explosion seems clear: the equation tells us that the energy inherent in any object is its mass multiplied by the square of the speed of light. Since the speed of light is huge, its square is even bigger, which equates to a lot of energy in a small amount of matter. The energy in one gram of matter turns out to be 90 terajoules, equivalent to about one day's output of electricity from a nuclear power station.

However, it didn't happen like that. The energy released in an atomic bomb is only a tiny fraction of the relativistic rest mass, and physicists were already aware, on experimental grounds, that certain nuclear reactions could release a lot of energy. The main technical problem was to hold a lump of suitable radioactive material together long enough to get a chain reaction, in which the decay of one radioactive atom causes it to emit radiation that triggers the same effect in other atoms, growing exponentially. Nevertheless, Einstein's equation quickly became established in the public mind as the progenitor of the atomic bomb. The Smyth report, an American government document released to the public to explain the atomic bomb, placed the equation on its second page. I suspect that what happened is what Jack Cohen and I have called 'lies to children' – simplified stories told for legitimate purposes, which pave the way to more accurate enlightenment.[3] This is how education works: the full story is always too complicated for anyone except the experts, and they know so much that they don't believe most of it.

However, Einstein's equation can't just be dismissed out of hand. It did play a role in the development of nuclear weapons. The notion of nuclear fission, which powers the atom bomb, arose from discussions between the physicists Lise Meitner and Otto Frisch in Nazi Germany in 1938. They were trying to understand the forces that held the atom together, which were a bit like the surface tension of a drop of liquid. They were out walking, discussing physics, and they applied Einstein's equation to work out whether fission was possible on energy grounds. Frisch later wrote:[4]

> We both sat down on a tree trunk and started to calculate on scraps of paper... When the two drops separated they would be driven apart by electrical repulsion, about 200 MeV in all. Fortunately Lise Meitner remembered how to compute the masses of nuclei... and worked out that the two nuclei formed... would be lighter by about one-fifth the mass of a proton... according to Einstein's formula $E = mc^2$... the mass was just equivalent to 200 MeV. It all fitted!

Although $E = mc^2$ was not directly responsible for the atom bomb, it was one of the big discoveries in physics that led to an effective theoretical understanding of nuclear reactions. Einstein's most important role regarding the atomic bomb was political. Urged by Leo Szilard, Einstein wrote to President Roosevelt, warning that the Nazis might be developing atomic weapons and explaining their awesome power. His reputation and influence were enormous, and the president heeded the warning. The Manhattan Project, Hiroshima and Nagasaki, and the ensuing Cold War were just some of the consequences.

Einstein wasn't satisfied with special relativity. It provided a unified theory of space, time, matter, and electromagnetism, but it missed out one vital ingredient.

Gravity.

Einstein believed that 'all the laws of physics' must satisfy his extended version of Galileo's principle of relativity. The law of gravitation surely ought to be among them. But that wasn't the case for the current version of relativity. Newton's inverse square law did not transform correctly between frames of reference. So Einstein decided he had to change Newton's law. He'd already changed virtually everything else in the Newtonian universe, so why not?

It took him ten years. His starting-point was to work out the implications of the principle of relativity for an observer moving freely under the influence of gravity – in a lift that is dropping freely, for example. Eventually he homed in on a suitable formulation. In this he was aided by a close friend, the mathematician Marcel Grossmann, who pointed him towards a rapidly growing field of mathematics: differential geometry. This had developed from Riemann's concept of a manifold and his characterisation of curvature, discussed in Chapter 1. There I mentioned that Riemann's metric can be written as a 3×3 matrix, and that technically this is a symmetric tensor. A school of Italian mathematicians, notably Tullio Levi-Civita and Gregorio Ricci-Curbastro, took up Riemann's ideas and developed them into tensor calculus.

From 1912, Einstein was convinced that the key to a relativistic theory of gravity required him to reformulate his ideas using tensor calculus, but in a 4-dimensional space-time rather than 3-dimensional space. The mathematicians were happily following Riemann and allowing any number of dimensions, so they had already set things up in more than enough generality. To cut a long story short, he eventually derived what we now call the Einstein field equations, which he wrote as:

$$R_{\mu\nu} - \tfrac{1}{2}Rg_{\mu\nu} = \kappa T_{\mu\nu}$$

Here R, g, and T are tensors – quantities that define physical properties and transform according to the rules of differential geometry – and κ is a constant. The subscripts μ and ν run over the four coordinates of space-time, so each tensor is a 4×4 table of 16 numbers. Both are symmetric, meaning that they don't change when μ and ν are swapped, which reduces them to a list of 10 distinct numbers. So really the formula packages together 10 equations, which is why we often refer to them using the plural – compare Maxwell's equations. R is Riemann's metric: it defines the shape of space-time. g is the Ricci curvature tensor, which is a modification of Riemann's notion of curvature. And T is the energy–momentum tensor, which describes how these two fundamental quantities depend on the space-time event concerned. Einstein presented his equations to the Prussian Academy of Science in 1915. He called his new work the general theory of relativity.

We can interpret Einstein's equations geometrically, and when we do, they provide a new approach to gravity. The basic innovation is that gravity is not represented as a force, but as the curvature of space-time. In the absence of gravity, space-time reduces to Minkowski space. The formula for the interval determines the corresponding curvature tensor. Its interpretation is 'not curved', just as Pythagoras's theorem applies to a flat plane but not to a positively or negatively curved non-Euclidean space. Minkowski space-time is flat. But when gravity occurs, space-time *bends*.

The usual way to picture this is to forget time, drop the dimensions of space down to two, and get something like Figure 51 (*left*). The flat plane of Minkowski space(-time) is distorted, shown here by an actual bend, creating a depression. Far from the star, matter or light travels in a straight line (dotted). But the curvature causes the path to bend. In fact, it looks superficially as though some force coming from the star attracts the matter towards it. But there is no force, just warped space-time. However, this image of curvature deforms space along an extra dimension, which is not required mathematically. An alternative image is to draw a grid of geodesics, shortest paths, equally spaced according to the curved metric. They bunch together where the curvature is greater, Figure 51 (*right*).

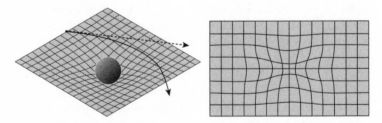

Fig 51 *Left*: Warped space near a star, and how it bends the paths of passing matter or light. *Right*: alternative image using a grid of geodesics, which bunch together in regions of higher curvature.

If the curvature of space-time is small, that is, if what (in the old picture) we think of as gravitational forces are not too large, then this formulation leads to Newton's law of gravity. Comparing the two theories, Einstein's constant κ turns out to be $8\pi G/c^4$, where G is Newton's gravitational constant. This links the new theory to the old one, and proves that in most cases the new one will agree with the old. The interesting new physics occurs when this is no longer true: when gravity is large. When Einstein came up with his theory, any test of relativity had to take place outside the laboratory, on a grand scale. Which meant astronomy.

Einstein therefore went looking for unexplained peculiarities in the motion of the planets, effects that didn't square with Newton. He found one that might be suitable: an obscure feature of the orbit of Mercury, the planet closest to the Sun, subjected to the greatest gravitational forces – and so, if Einstein was right, inside a region of high curvature.

Like all planets, Mercury follows a path that is very close to an ellipse, so some points in its orbit are closer to the Sun than others. The closest of all is called its *perihelion* ('near Sun' in Greek). The exact location of this perihelion had been observed for many years, and there was something funny about it. The perihelion slowly rotated about the Sun, an effect called precession; in effect, the long axis of the orbital ellipse was slowly changing direction. That was all right; Newton's laws predicted it, because Mercury is not the only planet in the Solar System and other planets were slowly changing its orbit. The problem was that Newtonian calculations gave the wrong rate of precession. The axis was rotating too quickly.

That had been known since 1840 when François Arago, director of the Paris Observatory, asked Urbain Le Verrier to calculate the orbit of Mercury using Newton's laws of motion and gravitation. But when the results were tested by observing the exact timing of a transit of Mercury – a passage across the face of the Sun, as viewed from Earth – they were wrong. Le Verrier decided to try again, eliminating potential sources of error, and in 1859 he published his new results. On the Newtonian model, the rate of precession was accurate to about 0.7%. The difference compared with observations was tiny: 38 seconds of arc every century (later revised to 43 arc-seconds). That's not much, less than one ten thousandth of a degree per year, but it was enough to interest Le Verrier. In 1846 he had made his reputation by analysing irregularities in the orbit of Uranus and predicting the existence, and location, of a then undiscovered planet: Neptune. Now he was hoping to repeat the feat. He interpreted the unexpected perihelion movement as evidence that some unknown world was perturbing Mercury's orbit. He did the sums and predicted the existence of a small planet with an orbit closer to the Sun than that of Mercury. He even had a name for it: Vulcan, the Roman god of fire.

Observing Vulcan, if it existed, would be difficult. The glare of the Sun was an obstacle, so the best bet was to catch Vulcan in transit, where it would be a tiny dark dot against the bright disc of the Sun. Shortly after

Le Verrier's prediction, an amateur astronomer named Edmond Lescarbault informed the distinguished astronomer that he had seen just that. He had initially assumed that the dot must be a sunspot, but it moved at the wrong speed. In 1860 Le Verrier announced the discovery of Vulcan to the Paris Academy of Science, and the government awarded Lescarbault the prestigious Légion d'Honneur.

Amid the clamour, some astronomers remained unimpressed. One was Emmanuel Liais, who had been studying the Sun with much better equipment than Lescarbault. His reputation was on the line: he had been observing the Sun for the Brazilian government, and it would have been disgraceful to have missed something of such importance. He flatly denied that a transit had taken place. For a time, everything got very confused. Amateurs repeatedly claimed they had seen Vulcan, sometimes years before Le Verrier announced his prediction. In 1878 James Watson, a professional, and Lewis Swift, an amateur, said they had seen a planet like Vulcan during a solar eclipse. Le Verrier had died a year earlier, still convinced he had discovered a new planet near the Sun, but without his enthusiastic new calculations of orbits and predictions of transits – none of which happened – interest in Vulcan quickly died away. Astronomers became skeptical.

In 1915, Einstein administered the *coup de grâce*. He reanalysed the motion using general relativity, without assuming any new planet, and a simple and transparent calculation led him to a value of 43 seconds of arc for the precession – the exact figure obtained by updating Le Verrier's original calculations. A modern Newtonian calculation predicts a precession of 5560 arc seconds per century, but observations give 5600. The difference is 40 seconds of arc, so about 3 arc-seconds per century remains unaccounted for. Einstein's announcement did two things: it was seen as a vindication of relativity, and as far as most astronomers were concerned, it relegated Vulcan to the scrapheap.[5]

Another famous astronomical verification of general relativity is Einstein's prediction that the Sun bends light. Newtonian gravitation also predicts this, but general relativity predicts an amount of bending that is twice as large. The total solar eclipse of 1919 provided an opportunity to distinguish the two, and Sir Arthur Eddington mounted an expedition, eventually announcing that Einstein prevailed. This was accepted with enthusiasm at the time, but later it became clear that the data were poor, and the result was questioned. Further independent observations from 1922 seemed to agree with the relativistic prediction, as did a later reanalysis of Eddington's data. By the 1960s it became possible to make the observations for radio-frequency radiation, and only then was it certain that the data did indeed show a deviation twice that predicted by Newton and equal to that predicted by Einstein.

The most dramatic predictions from general relativity arise on a much grander scale: black holes, which are born when a massive star collapses under its own gravitation, and the expanding universe, currently explained by the Big Bang.

Solutions to Einstein's equations are space-time geometries. These might represent the universe as a whole, or some part of it, assumed to be gravitationally isolated so that the rest of the universe has no important effect. This is analogous to early Newtonian assumptions that only two bodies are interacting, for example. Since Einstein's field equations involve ten variables, explicit solutions in terms of mathematical formulas are rare. Today we can solve the equations numerically, but that was a pipedream before the 1960s because computers either didn't exist or were too limited to be useful. The standard way to simplify equations is to invoke symmetry. Suppose that the initial conditions for a space-time are spherically symmetric, that is, all physical quantities depend only on the distance from the centre. Then the number of variables in any model is greatly reduced. In 1916 the German astrophysicist Karl Schwarzschild made this assumption for Einstein's equations, and managed to solve the resulting equations with an exact formula, known as the Schwarzschild metric. His formula had a curious feature: a singularity. The solution became infinite at a particular distance from the centre, called the Schwarzschild radius. At first it was assumed that this singularity was some kind of mathematical artefact, and its physical meaning was the subject of considerable dispute. We now interpret it as the event horizon of a black hole.

Imagine a star so massive that its radiation cannot counter its gravitational field. The star will begin to contract, sucked together by its own mass. The denser it gets, the stronger this effect becomes, so the contraction happens ever faster. The star's escape velocity, the speed with which an object must move to escape the gravitational field, also increases. The Schwarzschild metric tells us that at some stage, the escape velocity becomes equal to that of light. Now nothing can escape, because nothing can travel faster than light. The star has become a black hole, and the Schwarzschild radius tells us the region from which nothing can escape, bounded by the black hole's event horizon.

Black hole physics is complex, and there isn't space to do it justice here.

Suffice it to say that most cosmologists are now satisfied that the prediction is valid, that the universe contains innumerable black holes, and indeed that at least one lurks at the heart of our Galaxy. Indeed, of most galaxies.

In 1917 Einstein applied his equations to the entire universe, assuming another kind of symmetry: homogeneity. The universe should look the same (on large enough scale) at all points in space and time. By then he had modified the equations to include a 'cosmological constant' Λ, and sorted out the meaning of the constant κ. The equations were now written like this:

$$G_{\mu\nu} + \Lambda g_{\mu\nu} = \frac{8\pi G}{c^4 T_{\mu\nu}}$$

The solutions had a surprising implication: the universe should shrink as time passes. This forced Einstein to add the term involving the cosmological constant: he was seeking an unchanging, stable universe, and by adjusting the constant to the right value he could stop his model universe contracting to a point. In 1922 Alexander Friedmann found another equation, which predicted the universe should expand and did not require the cosmological constant. It also predicted the rate of expansion. Einstein still wasn't happy: he wanted the universe to be stable and unchanging.

For once Einstein's imagination failed him. In 1929 American astronomers Edwin Hubble and Milton Humason found evidence that the universe *is* expanding. Distant galaxies are moving away from us, as shown by shifts in the frequency of the light they emit – the famous Doppler effect, in which the sound of a speeding ambulance drops as it passes by, because the sound waves are affected by the relative speed of emitter and receiver. Now the waves are electromagnetic and the physics is relativistic, but there is still a Doppler effect. Not only do distant galaxies move away from us: the more distant they are, the faster they recede.

Running the expansion backwards in time, it turns out that at some point in the past, the entire universe was essentially just a point. Before that, it didn't exist at all. At that primeval point, space and time both came into existence in the famous Big Bang, a theory proposed by French mathematician Georges Lemaître in 1927, and almost universally ignored. When radio telescopes observed the cosmological microwave background radiation in 1964, at a temperature that fitted the Big Bang model, cosmologists decided Lemaître had been right after all. Again, the topic deserves a book of is own, and many have been written. Suffice it to say that our current most widely accepted theory of cosmology is an elaboration of the Big Bang scenario.

Scientific knowledge, however, is always provisional. New discoveries can change it. The Big Bang has been the accepted cosmological paradigm for the last 30 years, but it is beginning to show some cracks. Several discoveries either cast serious doubt on the theory, or require new physical particles and forces that have been inferred but not observed. There are three main sources of difficulty. I'll summarise them first, and then discuss each in more detail. The first is galactic rotation curves, which suggest that most of the matter in the universe is missing. The current proposal is that this is a sign of a new kind of matter, dark matter, which constitutes about 90% of the matter in the universe, and is different from any matter yet observed directly on Earth. The second is an acceleration in the expansion of the universe, which requires a new force, dark energy, of unknown origin but modelled using Einstein's cosmological constant. The third is a collection of theoretical issues related to the popular theory of inflation, which explains why the observable universe is so uniform. The theory fits observations, but its internal logic is looking shaky.

Dark matter first. In 1938 the Doppler effect was used to measure the speeds of galaxies in clusters, and the results were inconsistent with Newtonian gravitation. Because galaxies are separated by large distances, space-time is almost flat and Newtonian gravity is a good model. Fritz Zwicky suggested that there must be some unobserved matter to account for the discrepancy, and it was named dark matter because it could not be seen in photographs. In 1959, using the Doppler effect to measure the speed of rotation of stars in the galaxy M33, Louise Volders discovered that the observed rotation curve – a plot of speed against distance from the centre – was also inconsistent with Newtonian gravitation, which again is a good model. Instead of the speed falling off at greater distances, it remained almost constant, Figure 52. The same problem arises for many other galaxies.

If it exists, dark matter must be different from ordinary 'baryonic' matter, the particles observed in experiments on Earth. Its existence is accepted by most cosmologists, who argue that dark matter explains several different anomalies in observations, not just rotation curves. Candidate particles have been suggested, such as WIMPs (weakly interacting massive particles), but so far these particles have not been detected in experiments. The distribution of dark matter around galaxies

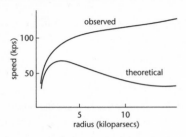

Fig 52 Galactic rotation curves for M33: theory and observations.

has been plotted by assuming dark matter exists and working out where it has to be to make the rotation curves flat. It generally seems to form two globes of galactic proportions, one above the plane of the galaxy and the other below it, like a giant dumb-bell. This is a bit like predicting the existence of Neptune from discrepancies in the orbit of Uranus, but such predictions require confirmation: Neptune had to be found.

Dark energy is similarly proposed to explain the results of the 1998 High-z Supernova Search Team, which expected to find evidence that the expansion of the universe is slowing down as the initial impulse from the Big Bang dies away. Instead, the observations indicated that the expansion of the universe is speeding up, a finding confirmed by the Supernova Cosmology Project in 1999. It is as though some antigravity force pervades space, pushing galaxies apart at an ever-increasing rate. This force is not any of the four basic forces of physics: gravity, electromagnetism, strong nuclear force, weak nuclear force. It was named dark energy. Again, its existence seemed to solve some other cosmological problems.

Inflation was proposed by the American physicist Alan Guth in 1980 to explain why the universe is extremely uniform in its physical properties on very large scales. Theory showed that the Big Bang ought to have produced a universe that was far more curved. Guth suggested that an 'inflaton field' (that's right, no second i: it's thought to be a scalar quantum field corresponding to a hypothetical particle, the inflaton) caused the early universe to expand with extreme rapidity. Between 10^{-36} and 10^{-32} seconds after the Big Bang, the volume of the universe grew by a mindboggling factor of 10^{78}. The inflaton field has not been observed (this would require unfeasibly high energies) but inflation explains so many features of the universe, and fits observations so closely, that most cosmologists are convinced it happened.

It's not surprising that dark matter, dark energy, and inflation were popular among cosmologists, because they let them continue to use their favourite physical models, and the results agreed with observations. But things are starting to fall apart.

The distributions of dark matter don't provide a satisfactory explanation of rotation curves. Enormous amounts of dark matter are needed to keep the rotation curve flat out to the large distances observed. The dark matter has to have unrealistically large angular momentum, which is inconsistent with the usual theories of galaxy formation. The same rather special initial distribution of dark matter is required in every galaxy, which seems unlikely. The dumb-bell shape is unstable because it places the additional mass on the outside of the galaxy.

Dark energy fares better, and it is thought to be some kind of quantum-mechanical vacuum energy, arising from fluctuations in the vacuum. However, current calculations of the size of the vacuum energy are too big by a factor of 10^{122}, which is bad news even by the standards of cosmology.[6]

The main problems affecting inflation are not observations – it fits those amazingly well – but its logical foundations. Most inflationary scenarios would lead to a universe that differs considerably from ours; what counts is the initial conditions at the time of the Big Bang. In order to match observations, inflation requires the early state of the universe to be very special. However, there are also very special initial conditions that produce a universe just like ours without invoking inflation. Although both sets of conditions are extremely rare, calculations performed by Roger Penrose[7] show that the initial conditions that do not require inflation outnumber those that produce inflation by a factor of one googolplex – ten to the power ten to the power 100. So explaining the current state of the universe without inflation would be much more convincing than explaining it with inflation.

Penrose's calculation relies on thermodynamics, which might not be an appropriate model, but an alternative approach, carried out by Gary Gibbons and Neil Turok, leads to the same conclusion. This is to 'unwind' the universe back to its initial state. It turns out that almost all of the potential initial states do not involve a period of inflation, and those that do require it are an exceedingly small proportion. But the biggest problem of all is that when inflation is wedded to quantum mechanics, it predicts that quantum fluctuations will occasionally trigger inflation in a small region of an apparently settled universe. Although such fluctuations are rare, inflation is so rapid and so gigantic that the net result is tiny islands of normal space-time surrounded by ever-growing regions of runaway inflation. In those regions, the fundamental constants of physics can be different from their values in our universe. In effect, anything is possible. Can a theory that predicts *anything* be testable scientifically?

There are alternatives, and it is starting to look as though they need to be taken seriously. Dark matter might not be another Neptune, but another Vulcan – an attempt to explain a gravitational anomaly by invoking new matter, when what really needs changing is the law of gravitation.

The main well-developed proposal is MOND, modified Newtonian dynamics, proposed by Israeli physicist Mordehai Milgrom in 1983. This modifies not the law of gravity, in fact, but Newton's second law of motion. It assumes that acceleration is not proportional to force when the acceleration is very small. There is a tendency among cosmologists to assume that the only viable alternative theories are dark matter or MOND – so if MOND disagrees with observations, that leaves only dark matter. However, there are many potential ways to modify the law of gravity, and we are unlikely to find the right one straight away. The demise of MOND has been proclaimed several times, but on further investigation no decisive flaw has yet been found. The main problem with MOND, to my mind, is that it puts into its equations what it hopes to get out; it's like Einstein modifying Newton's law to change the formula near a large mass. Instead, he found a radically new way to think of gravity, the curvature of space-time.

Even if we retain general relativity and its Newtonian approximation, there may be no need for dark energy. In 2009, using the mathematics of shock waves, American mathematicians Joel Smoller and Blake Temple showed that there are solutions of Einstein's field equations in which the metric expands at an accelerating rate.[8] These solutions show that small changes to the Standard Model could account for the observed acceleration of galaxies without invoking dark energy.

General relativity models of the universe assume that it forms a manifold; that is, on very large scales the structure smoothes out. However, the observed distribution of matter in the universe is clumpy on very big scales, such as the Sloan Great Wall, a filament composed of galaxies 1.37 billion light years long, Figure 53. Cosmologists believe that on even larger scales the smoothness will become apparent – but to date, every time the range of observations has been extended, the clumpiness has persisted.

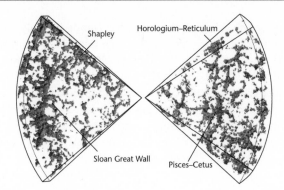

Fig 53 The clumpiness of the universe.

Robert MacKay and Colin Rourke, two British mathematicians, have argued that a clumpy universe in which there are many local sources of large curvature could explain all of the cosmological puzzles.[9] Such a structure is closer to what is observed than some large-scale smoothing, and is consistent with the general principle that the universe ought to be much the same everywhere. In such a universe there need be no Big Bang; in fact, the whole thing could be in a steady state, and be far, far older than the current figure of 13.8 billion years. Individual galaxies would go through a life cycle, surviving relatively unchanged for around 10^{16} years. They would have a very massive central black hole. Galactic rotation curves would be flat because of inertial drag, a consequence of general relativity in which a rotating massive body drags space-time round with it in its vicinity. The red shift observed in quasars would be caused by a large gravitational field, not by the Doppler effect, and would not be indicative of an expanding universe – this theory has long been advanced by American astronomer Halton Arp, and never satisfactorily disproved. The alternative model even indicates a temperature of $5°K$ for the cosmological microwave background, the main evidence (aside from red shift interpreted as expansion) for the Big Bang.

MacKay and Rourke say that their proposal 'overturns virtually every tenet of current cosmology. It does not, however, contradict any observational evidence.' It may well be wrong, but the fascinating point is that you can retain Einstein's field equations unchanged, dispense with dark matter, dark energy, and inflation, and *still* get behaviour reasonably like all of those puzzling observations. So whatever the theory's fate, it suggests that cosmologists should consider more imaginative mathematical models before resorting to new and otherwise unsupported physics. Dark matter, dark energy, inflation, each requiring radically new physics that no one has observed... In science, even one *deus ex machina* raises eyebrows. Three would be considered intolerable in anything other than cosmology. To be fair, it's difficult to experiment on the entire universe, so speculatively fitting theories to observations is about all that can be done. But imagine what would happen if a biologist explained life by some unobservable 'life field', let alone suggesting that a new kind of 'vital matter' and a new kind of 'vital energy' were also necessary – while providing no evidence that any of them existed.

Leaving aside the perplexing realm of cosmology, there are now more homely ways to verify relativity, both special and general, on a human scale. Special relativity can be tested in the laboratory, and modern measuring techniques provide exquisite accuracy. Particle accelerators such as the Large Hadron Collider simply would not work unless the designers took special relativity into account, because the particles that whirl round these machines do so at speeds very close indeed to that of light. Most tests of general relativity are still astronomical, ranging from gravitational lensing to pulsar dynamics, and the level of accuracy is high. A recent NASA experiment in low-Earth orbit, using high-precision gyroscopes, confirmed the occurrence of inertial frame-dragging, but failed to reach the intended precision because of unexpected electrostatic effects. By the time the data were corrected for this problem, other experiments had already achieved the same results.

However, one instance of relativistic dynamics, both special and general, is closer to home: car satellite navigation. The satnav systems used by motorists calculate the car's position using signals from a network of 24 orbiting satellites, the Global Positioning System. GPS is astonishingly accurate, and it works because modern electronics can reliably handle and measure very tiny instants of time. It is based on very precise timing signals, pulses emitted by the satellites and picked up on the ground. Comparing the signals from several satellites triangulates the location of the receiver to within a few metres. This level of accuracy requires knowing the timing to within about 25 nanoseconds (billionths of a second). Newtonian dynamics doesn't give correct locations, because two effects that are not accounted for in Newton's equations alter the flow of time: the satellite's motion and Earth's gravitational field.

Special relativity deals with the motion, and it predicts that the atomic clocks on the satellites should lose 7 microseconds (millionths of a second) per day compared with clocks on the ground, thanks to relativistic time dilation. General relativity predicts a gain of 45 microseconds per day caused by the Earth's gravity. The net result is that the clocks on the satellites gain 38 microseconds per day for relativistic reasons. Small as this may seem, its effect on GPS signals is by no means negligible. An error of 38 microseconds is 38,000 nanoseconds, about 1500 times the error that GPS can tolerate. If the software calculated your car's location using Newtonian dynamics, your satnav would quickly become useless, because the error would grow at a rate of 10 kilometres per day. Ten minutes from now Newtonian GPS would place you on the wrong street; by tomorrow it would place you in the wrong town. Within a week you'd be in the wrong county; within a month, the wrong country. Within a year, you'd be on the wrong planet. If you disbelieve relativity, but use satnav to plan your journeys, you have some explaining to do.

14 Quantum weirdness
Schrödinger's Equation

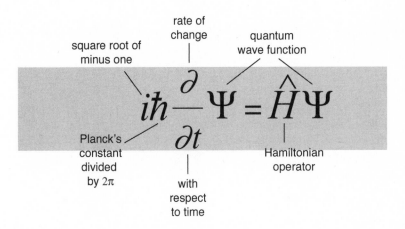

square root of
minus one

rate of
change

quantum
wave function

Planck's
constant
divided
by 2π

$$i\hbar \frac{\partial}{\partial t} \Psi = \hat{H}\Psi$$

with
respect
to time

Hamiltonian
operator

What does it say?

The equation models matter not as a particle, but as a wave, and describes how such a wave propagates.

Why is that important?

Schrödinger's equation is fundamental to quantum mechanics, which together with general relativity constitute today's most effective theories of the physical universe.

What did it lead to?

A radical revision of the physics of the world at very small scales, in which every object has a 'wave function' that describes a probability cloud of possible states. At this level the world is inherently uncertain. Attempts to relate the microscopic quantum world to our macroscopic classical world led to philosophical issues that still reverberate. But experimentally, quantum theory works beautifully, and today's computer chips and lasers wouldn't work without it.

n 1900 the great physicist Lord Kelvin declared that the then current theory of heat and light, considered to be an almost complete description of nature, was 'obscured by two clouds. The first involves the question: How could the Earth move through an elastic solid, such as is essentially the luminiferous ether? The second is the Maxwell–Boltzmann doctrine regarding the partition of energy.' Kelvin's nose for an important problem was spot on. In Chapter 13 we saw how the first question led to, and was resolved by, relativity. Now we will see how the second led to the other great pillar of present-day physics, quantum theory.

The quantum world is notoriously weird. Many physicists feel that if you don't appreciate just how weird it is, you don't appreciate it at all. There is a lot to be said for that opinion, because the quantum world is so different from our comfortable human-scale one that even the simplest concepts change out of all recognition. It is, for example, a world in which light is both a particle and a wave. It is a world where a cat in a box can be both alive and dead at the same time... until you open the box, that is, when suddenly the unfortunate animal's wave function 'collapses' to one state or the other. In the quantum multiverse, there exists one copy of our universe in which Hitler lost World War II, and another in which he won it. We just happen to live in – that is, exist as quantum wave functions in – the first one. Other versions of us, just as real but inaccessible to our senses, live in the other one.

Quantum mechanics is definitely weird. Whether it is quite *that* weird, though, is another matter altogether.

It all began with light bulbs. This was appropriate, because those were one of the most spectacular applications to emerge from the burgeoning subjects of electricity and magnetism, which Maxwell so brilliantly unified. In 1894 a German physicist named Max Planck was hired by an electrical company to design the most efficient light bulb possible, one giving the most light while consuming the least electrical energy. He saw that the key to this question was a fundamental issue in physics, raised in 1859 by another German physicist, Gustav Kirchhoff. It concerned a theoretical construct known as a black body, which absorbs all electromagnetic radiation that falls on it. The big question was: how does such a body *emit* radiation? It can't store it all; some has to come back out again. In particular, how does the intensity of the emitted radiation depend on its frequency and the body's temperature?

There was already an answer from thermodynamics, in which a black body can be modelled as a box whose walls are perfect mirrors. Electromagnetic radiation bounces to and fro, reflected by the mirrors. How is the energy in the box distributed among the various frequencies when the system has settled to an equilibrium state? In 1876 Boltzmann proved the 'equipartition theorem': the energy is apportioned equally to each independent component of the motion. These components are just like the basic waves in a violin string: normal modes.

There was only one problem with this answer: it couldn't possibly be correct. It implied that the total power radiated over all frequencies must be infinite. This paradoxical conclusion became known as the ultraviolet catastrophe: ultraviolet because that was the beginning of the high-frequency range, and catastrophe because it was. No real body can emit an infinite amount of power.

Although Planck was aware of this problem, it didn't bother him, because he didn't believe the equipartition theorem anyway. Ironically, his work resolved the paradox and did away with the ultraviolet catastrophe, but he noticed this only later. He used experimental observations of how energy depended on frequency, and fitted a mathematical formula to the data. His formula, derived early in 1900, did not initially have any physical basis. It was just a formula that worked. But later the same year he tried to reconcile his formula with the classical thermodynamic one, and decided that the energy levels of the black body's vibrational modes could not form a continuum, as thermodynamics assumed. Instead, these levels had to be discrete – separated by tiny gaps. In fact, for any given frequency, the energy had to be an integer multiple of that frequency, multiplied by a very tiny constant. We now call this number Planck's constant and denote it by h. Its value, in units of joule seconds, is $6.62606957(29) \times 10^{-34}$, where the figures in brackets may be inaccurate. This value is deduced from theoretical relationships between Planck's constant and other quantities that are easier to measure. The first such measurement was made by Robert Millikan using the photoelectric effect, described below. The tiny packets of energy are now called quanta (plural of quantum), from the Latin *quantus*, 'how much.'

Planck's constant may be tiny, but if the set of energy levels for a given frequency is discrete, the total energy turns out to be finite. So the ultraviolet catastrophe was a sign that a continuum model failed to reflect nature. And that implied that nature, on very small scales, must be discrete. Initially this didn't occur to Planck: he thought of his discrete energy levels as a mathematical trick to get a sensible formula. In fact, Boltzmann had entertained a similar idea in 1877, but didn't get anywhere with it. Everything changed when Einstein brought his fertile imagination to bear, and physics entered a new realm. In 1905, the same year as his work on special relativity, he investigated the photoelectric effect, in which light hitting a suitable metal causes it to emit electrons. Three years earlier Philipp Lenard had noticed that when the light has a higher frequency, the electrons have higher energies. But the wave theory of light, amply confirmed by Maxwell, implies that the energy of the electrons should depend on the intensity of the light, not on its frequency. Einstein realised that Planck's quanta would explain the discrepancy. He suggested that light, rather than being a wave, was composed of tiny particles, now called photons. The energy in a single photon, of a given frequency, should be the frequency times Planck's constant – just like one of Planck's quanta. A photon was a quantum of light.

There's an obvious problem with Einstein's theory of the photoelectric effect: it assumes light is a particle. But there was abundant evidence that light was a wave. On the other hand, the photoelectric effect was incompatible with light being a wave. So was light a wave, or a particle?

Yes.

It was – or had aspects that manifested themselves as – either. In some experiments, light seemed to behave like a wave. In others, it behaved like a particle. As physicists came to grips with very small scales of the universe, they decided that light wasn't the only thing to have this strange dual nature, sometimes particle, sometimes wave. All matter did. They called it wave–particle duality. The first person to grasp this dual nature of matter was Louis-Victor de Broglie, in 1924. He rephrased Planck's law in terms not of energy, but of momentum, and suggested that the momentum of the particle aspect and the frequency of the wave aspect should be related: multiply them together and you get Planck's constant. Three years later he was proved right, at least for electrons. One the one hand, electrons are particles, and can be observed behaving that way. On the other hand, they diffract like waves. In 1988 atoms of sodium were also spotted behaving like a wave.

Matter was neither particle nor wave, but a bit of both – a wavicle.

Several more or less intuitive images of this dual nature of matter were devised. In one, a particle is a localised clump of waves, known as a wave packet, Figure 54. The packet as a whole can behave like a particle, but some experiments can probe its internal wavelike structure. Attention shifted from providing images for wavicles to sorting out how they behaved. The quest quickly attained its goal, and the central equation of quantum theory emerged.

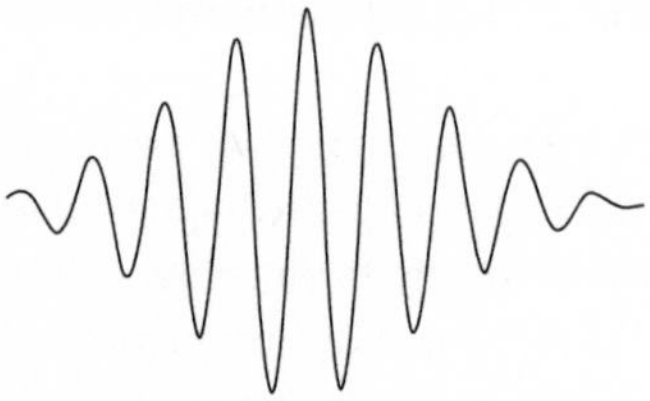

Fig 54 Wave packet.

The equation bears the name of Erwin Schrödinger. In 1927, building on the work of several other physicists, notably Werner Heisenberg, he wrote down a differential equation for any quantum wave function. It looked like this:

$$i\hbar \frac{\partial}{\partial t}\Psi = \hat{H}\Psi$$

Here Ψ (Greek capital psi) is the form of the wave, t is time (so $\partial/\partial t$ applied to Ψ gives its rate of change with respect to time), $\hat{H}$ is an expression called the Hamiltonian operator, and $\hbar$ is $h/2\pi$, where h is Planck's constant. And i? That was the weirdest feature of all. It's the square root of minus one (Chapter 5). Schrödinger's equation applies to waves defined over the *complex* numbers, not just the real numbers as in the familiar wave equation.

Waves in what? The classical wave equation (Chapter 8) defines waves in space, and its solution is a numerical function of space and time. The same goes for Schrödinger's equation, but now the wave function Ψ takes complex values, not just real ones. It's a bit like an ocean wave whose height is $2 + 3i$. The appearance of i is in many ways the most mysterious and profound feature of quantum mechanics. Previously i had turned up in solutions of equations, and in methods for finding those solutions, but here it was part of the equation, an explicit feature of the physical law.

One way to interpret this is that quantum waves are linked pairs of real waves, as if my complex ocean wave were really two waves, one of height 2 and the other of height 3, with the two directions of height at right angles to each other. But it's not quite that straightforward, because the two waves don't have a fixed shape. As time passes, they cycle through a whole series of shapes, and each is mysteriously linked to the other. It's a bit like the electric and magnetic components of a light wave, but now electricity can and does 'rotate' into magnetism, and conversely. The two waves are two facets of a single shape, which spins steadily around the unit circle in the complex plane. Both the real and the imaginary parts of this rotating shape change in a very specific way: they are combined in sinusoidally varying amounts. Mathematically this leads to the idea that a quantum wave function has a special kind of *phase*. The physical interpretation of that phase is similar to, but different from, the role of phase in the classical wave equation.

Remember how Fourier's trick solves both the heat equation and the wave equation? Some special solutions, Fourier's sines and cosines, have especially pleasant mathematical properties. All other solutions, however complicated, are superpositions of these normal modes. We can solve Schrödinger's equation using a similar idea, but now the basic patterns are more complicated than sines and cosines. They are called eigenfunctions, and they can be distinguished from all other solutions. Instead of being some general function of both space and time, an eigenfunction is a function defined only on space, multiplied by one depending only on time. The space and time variables, in the jargon, are separable. The eigenfunctions depend on the Hamiltonian operator, which is a mathematical description of the physical system concerned. Different systems – an electron in a potential well, a pair of colliding photons, whatever – have different Hamiltonian operators, hence different eigenfunctions.

For simplicity, consider a standing wave for the classical wave equation – a vibrating violin string, whose ends are pinned down. At all instants of time, the shape of the string is almost the same, but the amplitude is modulated: multiplied by a factor that varies sinusoidally with time, as in Figure 35 (page 138). The complex phase of a quantum wave function is similar, but harder to visualise. For any individual eigenfunction, the effect of the quantum phase is just a shift of the time coordinate. For a superposition of several eigenfunctions, you split the wave function into these components, factor each into a purely spatial part times a purely temporal one, spin the temporal part round the unit circle in the complex plane at the appropriate speed, and add the pieces back together. Each separate eigenfunction has a complex amplitude, and this modulates at its own particular frequency.

It may sound complicated, but it would be completely baffling if you didn't split the wave function into eigenfunctions. At least then you've got a chance.

Despite these complexities, quantum mechanics would be just a fancy version of the classical wave equation, resulting in two waves rather than one, were it not for a puzzling twist. You can observe classical waves, and see what shape they are, even if they are superpositions of several Fourier modes. But in quantum mechanics, you can never observe the entire wave function. All you can observe on any given occasion is a single component eigenfunction. Roughly speaking, if you attempt to measure two of these components at the same time, the measurement process on one of them disturbs the other one.

This immediately raises a difficult philosophical issue. If you can't observe the entire wave function, does it actually exist? Is it a genuine physical object, or just a convenient mathematical fiction? Is an unobservable quantity scientifically meaningful? It is here that Schrödinger's celebrated feline enters the story. It arises because of a standard way to interpret what a quantum measurement is, called the Copenhagen interpretation.[1]

Imagine a quantum system in some superposed state: say, an electron whose state is a mixture of spin-up and spin-down, which are pure states defined by eigenfunctions. (It doesn't matter what spin-up and spin-down mean.) When you observe the state, however, you either get spin-up, or you get spin-down. You can't observe a superposition. Moreover, once you've observed one of these – say spin-up – that *becomes* the actual state of the electron. Somehow your measurement seems to have forced the superposition to change into a specific component eigenfunction. This Copenhagen interpretation takes this statement literally: your measurement process has *collapsed* the original wave function into a single pure eigenfunction.

If you observe a lot of electrons, sometimes you get spin-up, sometimes spin-down. You can infer the probability that the electron is in one of those states. So the wave function itself can be interpreted as a kind of probability cloud. It doesn't show the actual state of the electron: it shows how probable it is that when you measure it, you get a particular result. But that makes it a statistical pattern, not a real *thing*. It no more proves the wave function is real than Quetelet's measurements of human height prove that a developing embryo possesses some sort of bell curve.

The Copenhagen interpretation is straightforward, reflects what happens in experiments, and makes no detailed assumptions about what happens when you observe a quantum system. For these reasons, most working physicists are very happy to use it. But some were not, in the early days when they theory was still being thrashed out, and some still are not. And one of the dissenters was Schrödinger himself.

In 1935, Schrödinger was worrying about the Copenhagen interpretation. He could see that it worked, on a pragmatic level, for quantum systems like electrons and photons. But the world around him, even though deep down inside it was just a seething mass of quantum particles, seemed different. Seeking a way to make the difference as glaring as he could, Schrödinger came up with a thought experiment in which a quantum particle had a dramatic and obvious effect on a cat.

Imagine a box, which when shut is impervious to all quantum interactions. Inside it, place an atom of radioactive matter, a radiation detector, a flask of poison, and a live cat. Now shut the box, and wait. At some point the radioactive atom will decay, and emit a particle of radiation. The detector will spot it, and is rigged so that when it does so, it causes the flask to break and release the poison inside. This kills the cat.

In quantum mechanics, the decay of a radioactive atom is a random event. From outside, no observer can tell whether the atom has decayed or not. If it has, the cat is dead; if not, it's alive. According to the Copenhagen interpretation, until someone observes the atom, it is in a superposition of two quantum states: decayed and not decayed. The same goes for the states of the detector, the flask, and the cat. So the cat is in a superposition of two states: dead and alive.

Since the box is impervious to all quantum interactions, the only way to find out whether the atom has decayed and killed the cat is to open the box. The Copenhagen interpretation tells us that the instant we do this, the wave functions collapse and the cat suddenly switches to a pure state: either dead, or alive. However, the inside of the box is no different from the external world, where we never observe a cat that is in a superposed alive/ dead state. So before we open the box and observe its contents, there must either be a dead cat inside, or a live one.

Schrödinger intended this thought experiment as a criticism of the Copenhagen interpretation. Microscopic quantum systems obey the superposition principle and can exist in mixed states; macroscopic ones can't. By linking a microscopic system, the atom, to a macroscopic one, the cat, Schrödinger was pointing out what he believed to be a flaw in the Copenhagen interpretation: it gave nonsense when applied to a cat. He must have been startled when the majority of physicists responded, in effect: 'Yes, Erwin, you're absolutely right: until someone opens the box, the cat really is simultaneously dead and alive.' Especially when it dawned on him that he couldn't find out who was right, even if he opened the box. He would observe either a live cat or a dead one. He might infer that the cat had been in that state before he opened the box, but he couldn't be sure. The observable result was consistent with the Copenhagen interpretation.

Very well: add a film camera to the contents of the box, and film what actually happens. That will decide the matter. 'Ah, no,' the physicists replied. 'You can only see what the camera has filmed after you open the box. Before that, the film is in a superposed state: containing a movie of a live cat, and containing a movie of a dead one.'

The Copenhagen interpretation freed up physicists to do their calculations and sort out what quantum mechanics predicted, without facing up to the difficult, if not impossible, issue of how the classical world emerged from a quantum substrate – how a macroscopic device, unimaginably complex on a quantum scale, somehow made a measurement of a quantum state. Since the Copenhagen interpretation did the job, they weren't really interested in philosophical questions. So generations of physicists were taught that Schrödinger had invented his cat to show that quantum superposition extended into the macroscopic world too: the exact opposite of what Schrödinger had been trying to tell them.

It's not really a great surprise that matter behaves strangely on the level of electrons and atoms. We may initially rebel at the thought, out of unfamiliarity, but if an electron is really a tiny clump of waves rather than a tiny clump of *stuff*, we can learn to live with it. If that means that the state of the electron is itself a bit weird, spinning not just about an up axis or a down axis but a bit of both, we can live with that too. And if the limitations of our measuring devices imply that we can never catch the electron doing that kind of thing – that any measurement we make necessarily settles for some pure state, up or down – then that's how it is. If the same applies to a radioactive atom, and the states are 'decayed' or 'not decayed', because its component particles have states as elusive as those of the electron, we can even accept that the atom itself, in its entirety, may be in a superposition of those states until we make a measurement. But a cat is a cat, and it seems to be a very big stretch of the imagination to imagine that the animal can be both alive and dead at the same time, only to miraculously collapse into one or the other when we open the box that contains it. If quantum reality requires a superposed alive/dead cat, why is it so shy that it won't let us observe such a state?

There are sound reasons in the formalism of quantum theory that (until very recently) require any measurement, any 'observable', to be an eigenfunction. There are even sounder reasons why the state of a quantum system should be a wave, obeying Schrödinger's equation. How can you get from one to the other? The Copenhagen interpretation declares that somehow (don't ask how) the measurement process collapses the complex, superposed wave function down to a single component eigenfunction. Having been provided with this form of words, your task as a physicist is to get on with making measurements and calculating eigenfunctions and so on, and stop asking awkward questions. It works amazingly well, if you measure success by getting answers that agree with experiment. And everything would have been fine if Schrödinger's equation permitted the wave function to behave in this manner, but it doesn't. In *The Hidden Reality* Brian Greene puts it this way: 'Even polite prodding reveals an uncomfortable feature... The instantaneous collapse of a wave... can't possible emerge from Schrödinger's math.' Instead, the Copenhagen interpretation was a pragmatic bolt-on to the theory, a way to handle measurements without understanding or facing up to what they really were.

This is all very well, but it's not what Schrödinger was trying to point out. He introduced a cat, rather than an electron or an atom, because it put what he considered to be the main issue in sharp relief. A cat belongs to the macroscopic world in which we live, in which matter does not behave the way quantum mechanics demands. We do not see superposed cats.[2] Schrödinger was asking why our familiar 'classical' universe fails to resemble the underlying quantum reality. If everything from which the world is built can exist in superposed states, why does the universe look classical? Many physicists have performed wonderful experiments showing that electrons and atoms really do behave the way quantum and

Copenhagen say they should. But this misses the point: you have to do it with a cat. Theorists wondered whether the cat could observe its own state, or whether someone else could secretly open the box and write down what was inside. They concluded, following the same logic as Schrödinger, that if the cat observed its state then the box contained a superposition of a dead cat that had committed suicide by observing itself, and a live cat that had observed itself to be alive, until the legitimate observer (a physicist) opened the box. Then the whole shebang collapsed to one or the other. Similarly the friend became a superposition of two friends: one of whom had seen a dead cat while the other had seen a live one, until a physicist opened the box, causing the friend's state to collapse. You could proceed in this way until the state of the entire *universe* was a superposition of a universe with a dead cat and a universe with a live one, and then the state of the universe collapsed when a physicist opened the box.

It was all a bit embarrassing. Physicists could get on with their work without sorting it out, they could even deny there was anything to *be* sorted out, but something was missing. For example, what happens to us if an alien physicist on the planet Apellobetnees III opens a box? Do we suddenly discover we actually blew ourselves up in a nuclear war when the Cuban missile crisis of 1962 escalated, and have been living on borrowed time ever since?

The measurement process is not the neat, tidy mathematical operation that the Copenhagen interpretation assumes. When asked to describe how the apparatus comes to its decision, the Copenhagen interpretation replies 'it just does'. The image of the wave function collapsing to a single eigenfunction describes the input and the output of the measurement process, but not how to get from one to the other. But when you make a real measurement you don't just wave a magic wand and cause the wave function to disobey Schrödinger's equation and collapse. Instead, you do something so enormously complicated, from a quantum viewpoint, that it is obviously hopeless to model it realistically. To measure an electron's spin, for example, you make it interact with a suitable piece of apparatus, which has a pointer that either moves to the 'up' position or the 'down' one. Or a numerical display, or a signal sent to a computer... This device yields *one* state, and one state only. You don't see the pointer in a superposition of up and down.

We are used to this, because that's how the classical world works. But underneath it's supposed to be a quantum world. Replace the cat with the spin apparatus, and it should indeed exist in a superposed state. The apparatus, viewed as a quantum system, is extraordinarily complicated. It contains gazillions of particles – between 10^{25} and 10^{30}, at a rough estimate. The measurement emerges somehow from the interaction of that single electron with these gazillion particles. Admiration for the expertise of the company that manufactures the instrument must be boundless; to extract anything sensible from something so messy is almost beyond belief. It's like trying to work out someone's shoe size by making them pass through a city. But if you're clever (arrange for them to encounter a shoe shop) you can get a sensible result, and a clever instrument designer can produce meaningful measurements of electron spin. But there's no realistic prospect of modelling in detail how such a device works as a *bona fide* quantum system. There's too much detail, the biggest computer in the world would flounder. That makes it difficult to analyse a real measurement process using Schrödinger's equation.

Even so, we do have some understanding of how our classical world emerges from an underlying quantum one. Let's start with a simple version, a ray of light hitting a mirror. The classical answer, Snell's law, states that the reflected ray bounces off at the same angle as the one that hit. In his book *QED* on quantum electrodynamics, the physicist Richard Feynman explained that this is not what happens in the quantum world. The ray is really a stream of photons, and each photon can bounce all over the place. However, if you superpose all the possible things the photon could do, then you get Snell's law. The overwhelming proportion of photons bounce back at angles very close to the one at which they hit. Feynman even managed to show why without using any complicated mathematics, but behind this calculation is a general mathematical idea: the principle of stationary phase. If you superpose all quantum states for an optical system, you get the classical outcome in which light rays follow the shortest path, measured by time taken. You can even add bells and whistles to decorate the ray paths with classical wave-optical diffraction fringes.

This example shows, very explicitly, that the superposition of all possible worlds – in this optical framework – yields the classical world. The most important feature is not so much the detailed geometry of the light ray, but the fact that it yields only *one* world at the classical level. Down in the quantum details of individual photons, you can observe all the paraphernalia of superposition, eigenfunctions, and so on. But up at the human scale, all that cancels out – well, adds together – to produce a clean, classical world.

The other part of the explanation is called decoherence. We've seen that quantum waves have a phase as well as an amplitude. It's a very funny phase, a complex number, but it's a phase nonetheless. The phase is absolutely crucial to any superposition. If you take two superposed states, change the phase of one, and add them back together, what you get is nothing like the original. If you do the same with a lot of components, the reassembled wave can be almost anything. Loss of phase information wrecks any Schrödinger's cat-like superposition. You don't just lose track of whether it's alive or dead: you can't tell it was a cat. When quantum waves cease to have nice phase relations, they decohere – they start to behave more like classical physics, and superpositions lose any meaning. What causes them to decohere is interactions with surrounding particles. This is presumably how apparatus can measure electron spin and get a specific, unique result.

Both of these approaches lead to the same conclusion: classical physics is what you observe if you take a human-scale view of a very complicated quantum system with gazillions of particles. Special experimental methods, special devices, might preserve some of the quantum effects, making them poke up into our comfortable classical existence, but generic quantum systems quickly cease to appear quantum as we move to larger scales of behaviour.

That's one way to resolve the fate of the poor cat. Only if the box is totally impervious to quantum decoherence can the experiment produce the superposed cat, and *no such box exists*. What would you make it from?

But there's another way, one that goes to the opposite extreme. Earlier I said that 'You could proceed in this way until the state of the entire *universe* was a superposition.' In 1957 Hugh Everett Jr. pointed out that in a sense, you have to. The only way to provide an accurate quantum model of a system is to consider its wave function. Everyone was happy to do so when the system was an electron, or an atom, or (more controversially) a cat. Everett took the system to be the entire universe.

He argued that you had no choice if that's what you wanted to model. Nothing less than the universe can be truly isolated. Everything interacts with everything else. And he discovered that if you took that step, then the problem of the cat, and the paradoxical relation between quantum and classical reality, is easily resolved. The quantum wave function of the universe is not a pure eigenmode, but a superposition of all possible eigenmodes. Although we can't calculate such things (we can't for a cat, and a universe is a tad more complicated) we can reason about them. In effect, we are representing the universe, quantum-mechanically, as a combination of *all the possible things that a universe can do.*

The upshot was that the wave function of the cat does not have to collapse to give a single classical observation. It can remain completely unchanged, with no violation of Schrödinger's equation. Instead, there are two coexisting universes. In one, the cat died; in the other, it didn't. When you open the box, there are correspondingly two yous and two boxes. One of them is part of the wave function of a universe with a dead cat; the other is part of a different wave function with a live cat. In place of a unique classical world that somehow emerges from the superposition of quantum possibilities, we have a vast range of classical worlds, each corresponding to one quantum possibility.

Everett's original version, which he called the relative state formulation, came to popular attention in the 1970s through Bryce DeWitt, who gave it a more catchy name: the many-worlds interpretation of quantum mechanics. It is often dramatised in historical terms: for example, that there is a universe in which Adolf Hitler won World War II, and another one in which he didn't. The one in which I am writing this book is the latter, but somewhere alongside it in the quantum realm another Ian Stewart is writing a book very similar to this one, but in German, reminding his readers that they are in the universe where Hitler won. Mathematically, Everett's interpretation can be viewed as a logical equivalent of conventional quantum mechanics, and it leads – in more limited interpretations – to efficient ways to solve physics problems. His formalism will therefore survive any experimental test that conventional quantum mechanics survives. So does that imply that these parallel universes, 'alternate worlds' in transatlantic parlance, *really* exist? Is another me typing away happily on a computer keyboard in a world where Hitler won? Or is the set-up a convenient mathematical fiction?

There is an obvious problem: how can we be sure that in a world dominated by Hitler's dream, the Thousand Year Reich, computers like the one I'm using would exist? Clearly there must be many more universes than two, and events in them must follow sensible classical patterns. So maybe Stewart-2 doesn't exist but Hitler-2 does. A common description of the formation and evolution of parallel universes involves them 'splitting off' whenever there is a choice of quantum state. Greene points out that this image is wrong: nothing splits. The universe's wave function has been, and always will be, split. Its component eigenfunctions are *there*: we imagine a split when we select one of them, but the whole point of

Everett's explanation is that nothing in the wave function actually changes.

With that as a caveat, a surprising number of quantum physicists accept the many-worlds interpretation. Schrödinger's cat really is alive and dead. Hitler really did win and lose. One version of us lives in one of those universes, others do not. That's what the mathematics says. It's not an interpretation, a convenient way to arrange the calculations. It's as real as you and I. It *is* you and I.

I'm not convinced. It's not the superposition that bothers me, though. I don't find the existence of a parallel Nazi world unthinkable, or impossible.[3] But I do object, strenuously, to the idea that you can separate a quantum wave function according to human-scale historical narratives. The mathematical separation occurs at the level of quantum states of constituent particles. Most combinations of particle states make no sense whatsoever as a human narrative. A simple alternative to a dead cat is not a live cat. It is a dead cat with one electron in a different state. Complex alternatives are far more numerous than a live cat. They include a cat that suddenly explodes for no apparent reason, one that turns into a flower vase, one that gets elected president of the United States, and one that survived even though the radioactive atom released the poison. Those alternative cats are rhetorically useful but unrepresentative. Most alternatives are not cats at all; in fact, *they are indescribable in classical terms.* If so, most of the alternative Stewarts aren't recognisable as people – indeed as anything – and almost all of those that exist do so within a world that makes absolutely no sense in human terms. So the chance that another version of little old me happens to live in another world that makes narrative sense to a human being is negligible.

The universe may well be an incredibly complex superposition of alternative states. If you think quantum mechanics is basically right, it has to be. In 1983 the physicist Stephen Hawking said that the many-worlds interpretation was 'self-evidently correct' in this sense. But it doesn't follow that there exists a superposition of universes in which a cat is alive or dead, and Hitler did or did not win. There is no reason to suppose that the mathematical components can be separated into sets that fit together to create human narratives. Hawking dismissed narrative interpretations of the many-worlds formalism, saying 'All that one does, really, is to calculate conditional probabilities – in other words, the probability of A happening, given B. I think that that's all the many-worlds interpretation is. Some people overlay it with a lot of mysticism about the wave function splitting into different parts. But all that you're calculating is conditional probabilities.'

It's worth comparing the Hitler tale with Feynman's story of the light ray. In the style of alternative Hitlers, Feynman would be telling us that there is one classical world where the light ray bounces off the mirror at the same angle at which it hit, another classical world in which it bounces at an angle that's one degree wrong, another where it's two degrees wrong, and so on. But he didn't. He told us that there is *one* classical world, emerging from the superposition of the quantum alternatives. There may be innumerable parallel worlds at the quantum level, but these do not correspond in any meaningful way to parallel worlds that are describable at the classical level. Snell's law is valid in *any* classical world. If it weren't, the world couldn't be classical. As Feynman explained for light rays, *the* classical world emerges when you superpose all of the quantum alternatives. There is only one such superposition, so there is only one classical universe. Ours.

Quantum mechanics isn't confined to the laboratory. The whole of modern electronics depends on it. Semiconductor technology, the basis of all integrated circuits – silicon chips – is quantum-mechanical. Without the physics of the quantum, no one would have dreamed that such devices could work. Computers, mobile phones, CD players, games consoles, cars, refrigerators, ovens, virtually all modern household gadgets, contain memory chips, to contain the instructions that make these devices do what we want. Many contain more complex circuitry, such as microprocessors, an entire computer on a chip. Most memory chips are variations on the first true semiconductor device: the transistor.

In the 1930s, American physicists Eugene Wigner and Frederick Seitz analysed how electrons move though a crystal, a problem that requires quantum mechanics. They discovered some of the basic features of semiconductors. Some materials are conductors of electricity: electrons can flow through them with ease. Metals are good conductors, and in everyday use copper wire is commonplace for this purpose. Insulators do not permit electrons to flow, so they stop the flow of electricity: the plastics that sheathe electrical wires, to prevent us electrocuting ourselves on the TV power lead, are insulators. Semiconductors are a bit of both, depending on circumstances. Silicon is the best known, and currently the most widely used, but several other elements such as antimony, arsenic, boron, carbon, germanium, and selenium are also semiconductors. Because semiconductors can be switched from one state to the other, they can be used to manipulate electrical currents, and this is the basis of all electronic circuits.

Wigner and Seitz discovered that the properties of semiconductors depend on the energy levels of the electrons within them, and these levels can be controlled by 'doping' the basic semiconductor material by adding small quantities of specific impurities. Two important types are p-type semiconductors, which carry current as a flow of electrons, and n-type semiconductors, in which current flows in the opposite direction to the electrons, carried by 'holes' – places where there are fewer electrons than normal. In 1947 John Bardeen and Walter Brattain at Bell Laboratories discovered that a crystal of germanium could act as an amplifier. If an electrical current was fed into it, the output current was higher. William Shockley, leader of the Solid State Physics Group, realised how important this could be, and initiated a project to investigate semiconductors. Out of this came the transistor – short for 'transfer resistor'. There were some earlier patents but no working devices or published papers. Technically the Bell Labs' device was a JFET (junction gate field-effect transistor, Figure 55). Since this initial breakthrough, many other kinds of transistor have been invented. Texas Instruments manufactured the first silicon transistor in 1954. The same year saw a transistor-based computer, TRIDAC, built by the US military. It was three cubic feet in size and its power requirement was the same as one light bulb. This was an early step in a huge American military programme to develop alternatives to vacuum tube electronics, which was too cumbersome, fragile, and unreliable for military use.

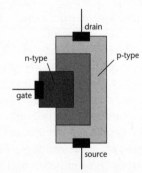

Fig 55 Structure of a JFET. The source and drain are at the ends, in a p-type layer, while the gate is an n-type layer that controls the flow. If you think of the flow of electrons from source to drain as a hose, the gate in effect squeezes the hose, increasing the pressure (voltage) at the drain.

Because semiconductor technology is based on doping silicon or similar substances with impurities, it lent itself to miniaturisation. Circuits can be built up in layers on a silicon substrate, by bombarding the surface with the desired impurity, and etching away unwanted regions with acid. The areas affected are determined by photographically produced masks, and these can be shrunk to very small size using optical lenses. Out of all this emerged today's electronics, including memory chips that can hold billions of bytes of information and very fast microprocessors that orchestrate the activity of computers.

Another ubiquitous application of quantum mechanics is the laser. This is a device that emits a strong beam of coherent light: one in which the light waves are all in phase with each other. It consists of an optical cavity with mirrors at each end, filled with something that reacts to light of a specific wavelength by producing more light of the same wavelength – a light amplifier. Pump in energy to start the process rolling, let the light bounce to and fro along the cavity, amplifying all the time, and when it reaches a sufficiently high intensity, let it out. The gain medium can be a fluid, a gas, a crystal, or a semiconductor. Different materials work at different wavelengths. The amplification process depends on the quantum mechanics of atoms. The electrons in the atoms can exist in different energy states, and they can be switched between them by absorbing or emitting photons.

LASER means light amplification by stimulated emission of radiation. When the first laser was invented, it was widely derided as an answer looking for a problem. This was unimaginative: a whole host of suitable problems quickly appeared, once there was a solution. Producing a coherent beam of light is basic technology, and it was always bound to have uses, just as an improved hammer would automatically find many uses. When inventing generic technology, you don't have to have a specific application in mind. Today we use lasers for so many purposes that it's impossible to list them all. There are prosaic uses like laser pointers for lectures and laser beams for DIY. CD players, DVD players, and Blu-ray all use lasers to read information from tiny pits or marks on discs. Surveyors use lasers to measure distances and angles. Astronomers use lasers to measure the distance from the Earth to the Moon. Surgeons use lasers for fine cutting of delicate tissues. Laser treatment of eyes is routine, for repairing detached retinas and remoulding the surface of the cornea to correct vision instead of using glasses or contact lenses. The 'Star Wars' antimissile system was intended to use powerful lasers to shoot down enemy missiles, and although it was never built, some of the lasers were. Military uses of lasers, akin to the pulp science-fiction ray-gun, are being investigated right now. And it may even be possible to launch space vehicles from Earth by making them ride a powerful laser beam.

New uses of quantum mechanics arrive almost by the week. One of the latest is quantum dots, tiny pieces of semiconductor whose electronic properties, including the light that they emit, vary according to their size and shape. They can therefore be tailored to have many desirable features. They already have a variety of applications, including biological imaging, where they can replace traditional (and often toxic) dyes. They also perform much better, emitting brighter light.

Further down the line, some engineers and physicists are working on the basic components of a quantum computer. In such a device, the binary states of 0 and 1 can be superposed in any combination, in effect allowing computations to assume both values at the same time. This would allow many different calculations to be performed in parallel, speeding them up enormously. Theoretical algorithms have been devised, carrying out such tasks as splitting a number into its prime factors. Conventional computers run into trouble when the numbers have more than a hundred digits or so, but a quantum computer should be able to factorise much bigger numbers with ease. The main obstacle to quantum computing is decoherence, which destroys superposed states. Schrödinger's cat is exacting revenge for its inhumane treatment.

15 Codes, communications, and computers
Information Theory

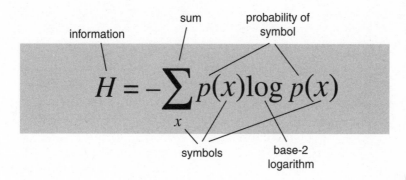

What does it say?

It defines how much information a message contains, in terms of the probabilities with which the symbols that make it up are likely to occur.

Why is that important?

It is the equation that ushered in the information age. It established limits on the efficiency of communications, allowing engineers to stop looking for codes that were too effective to exist. It is basic to today's digital communications – phones, CDs, DVDs, the internet.

What did it lead to?

Efficient error-detecting and error-correcting codes, used in everything from CDs to space probes. Applications include statistics, artificial intelligence, cryptography, and extracting meaning from DNA sequences.

n 1977 NASA launched two space probes, *Voyager 1* and *2*. The planets of the Solar System had arranged themselves in unusually favourable positions, making it possible to find reasonably efficient orbits that would let the probes visit several planets. The initial aim was to examine Jupiter and Saturn, but if the probes held out, their trajectories would take them on past Uranus and Neptune. *Voyager 1* could have gone to Pluto (at that time considered a planet, and equally interesting – indeed totally unchanged – now that it's not) but an alternative, Saturn's intriguing moon Titan, took precedence. Both probes were spectacularly successful, and *Voyager 1* is now the most distant human-made object from Earth, more than 10 billion miles away and still sending back data.

Signal strength falls off with the square of the distance, so the signal received on Earth is 10^{-20} times the strength that it would be if received from a distance of one mile. That is, one hundred quintillion times weaker. *Voyager 1* must have a really powerful transmitter... No, it's a tiny space probe. It is powered by a radioactive isotope, plutonium-238, but even so the total power available is now about one eighth that of a typical electric kettle. There are two reasons why we can still obtain useful information from the probe: powerful receivers on Earth, and special codes used to protect the data from errors caused by extraneous factors such as interference.

Voyager 1 can send data using two different systems. One, the low-rate channel, can send 40 binary digits, 0s or 1s, every second, but it does not allow coding to deal with potential errors. The other, the high-rate channel, can transmit up to 120,000 binary digits every second, and these are encoded so that errors can be spotted and put right provided they're not too frequent. The price paid for this ability is that the messages are twice as long as they would otherwise be, so they carry only half as much data as they could. Since errors could ruin the data, this is a price worth paying.

Codes of this kind are widely used in all modern communications: space missions, landline phones, mobile phones, the Internet, CDs and DVDs, Blu-ray, and so on. Without them, all communications would be liable to errors; this would not be acceptable if, for instance, you were using the Internet to pay a bill. If your instruction to pay £20 was received as £200, you wouldn't be pleased. A CD player uses a tiny lens, which focuses a laser beam on to very thin tracks impressed in the material of the disc. The lens hovers a very tiny distance above the spinning disc. Yet you can listen to a CD while driving along a bumpy road, because the signal is encoded in a way that allows errors to be found and put right by the electronics while the disc is being played. There are other tricks, too, but this one is fundamental.

Our information age relies on digital signals – long strings of 0s and 1s, pulses and non-pulses of electricity or radio. The equipment that sends, receives, and stores the signals relies on very small, very precise electronic circuits on tiny slivers of silicon – 'chips'. But for all the cleverness of the circuit design and manufacture, none of it would work without error-detecting and error-correcting codes. And it was in this context that the term 'information' ceased to be an informal word for 'know-how', and became a measurable numerical quantity. And that provided fundamental limitations on the efficiency with which codes can modify messages to protect them against errors. Knowing these limitations saved engineers lots of wasted time, trying to invent codes that would be so efficient they'd be impossible. It provided the basis for today's information culture.

I'm old enough to remember when the only way to telephone someone in *another country* (shock horror) was to make a booking ahead of time with the phone company – in the UK there was only one, Post Office Telephones – for a specific time and length. Say ten minutes at 3.45 p.m. on 11 January. And it cost a fortune. A few weeks ago a friend and I did an hour-long interview for a science fiction convention in Australia, from the United Kingdom, using Skype^{TM}. It was free, and it sent video as well as sound. A lot has changed in fifty years. Nowadays, we exchange information online with friends, both real ones and the phonies that large numbers of people collect like butterflies using social networking sites. We no longer buy music CDs or movie DVDs: we buy the information that they contain, downloaded over the Internet. Books are heading the same way. Market research companies amass huge quantities of information about our purchasing habits and try to use it to influence what we buy. Even in medicine, there is a growing emphasis on the information that is contained in our DNA. Often the attitude seems to be that if you have the information required to do something, then that alone suffices; you don't need actually to do it, or even know how to do it.

There is little doubt that the information revolution has transformed our lives, and a good case can be made that in broad terms the benefits outweigh the disadvantages – even though the latter include loss of privacy, potential fraudulent access to our bank accounts from anywhere in the word at the click of a mouse, and computer viruses that can disable a bank or a nuclear power station.

What is information? Why does it have such power? And is it really what it is claimed to be?

The concept of information as a measurable quantity emerged from the research laboratories of the Bell Telephone Company, the main provider of telephone services in the United States from 1877 to its break-up in 1984 on anti-trust (monopoly) grounds. Among its engineers was Claude Shannon, a distant cousin of the famous inventor Edison. Shannon's best subject at school was mathematics, and he had an aptitude for building mechanical devices. By the time he was working for Bell Labs he was a mathematician and cryptographer, as well as an electronic engineer. He was one of the first to apply mathematical logic – so-called Boolean algebra – to computer circuits. He used this technique to simplify the design of switching circuits used by the telephone system, and then extended it to other problems in circuit design.

During World War II he worked on secret codes and communications, and developed some fundamental ideas that were reported in a classified memorandum for Bell in 1945 under the title 'A mathematical theory of cryptography'. In 1948 he published some of his work in the open literature, and the 1945 article, declassified, was published soon after. With additional material by Warren Weaver, it appeared in 1949 as *The Mathematical Theory of Communication*.

Shannon wanted to know how to transmit messages effectively when the transmission channel was subject to random errors, 'noise' in the engineering jargon. All practical communications suffer from noise, be it from faulty equipment, cosmic rays, or unavoidable variability in circuit components. One solution is to reduce the noise by building better equipment, if possible. An alternative is to encode the signals using mathematical procedures that can detect errors, and even put them right.

The simplest error-detecting code is to send the same message twice. If you receive

the same massage twice
the same message twice

then there is clearly an error in the third word, but without understanding English it is not obvious which version is correct. A third repetition would decide the matter by majority vote and become an error-correcting code. How effective or accurate such codes are depends on the likelihood, and nature, of the errors. If the communication channel is very noisy, for instance, then all three versions of the message might be so badly garbled that it would be impossible to reconstruct it.

In practice mere repetition is too simple-minded: there are more efficient ways to encode messages to reveal or correct errors. Shannon's starting point was to pinpoint the meaning of efficiency. All such codes replace the original message by a longer one. The two codes above double or treble the length. Longer messages take more time to send, cost more, occupy more memory, and clog the communication channel. So the efficiency, for a given rate of error detection or correction, can be quantified as the ratio of the length of the coded message to that of the original.

The main issue, for Shannon, was to determine the inherent limitations of such codes. Suppose an engineer had devised a new code. Was there some way to decide whether it was about as good as they get, or might some improvement be possible? Shannon began by quantifying how much information a message contains. By so doing, he turned 'information' from a vague metaphor into a scientific concept.

There are two distinct ways to represent a number. It can be defined by a sequence of symbols, for example its decimal digits, or it can correspond to some physical quantity, such as the length of a stick or the voltage in a wire. Representations of the first kind are digital, the second are analogue. In the 1930s, scientific and engineering calculations were often performed using analogue computers, because at the time these were easier to design and build. Simple electronic circuits can add or multiply two voltages, for example. However, machines of this type lacked precision, and digital computers began to appear. It very quickly became clear that the most convenient representation of numbers was not decimal, base 10, but binary, base 2. In decimal notation there are ten symbols for digits, 0–9, and every digit multiplies ten times in value for every step it moves to the left. So 157 represents

$$1 \times 10^2 + 5 \times 10^1 + 7 \times 10^0$$

Binary notation employs the same basic principle, but now there are only two digits, 0 and 1. A binary number such as 10011101 encodes, in symbolic form, the number

$$1 \times 2^7 + 0 \times 2^6 + 0 \times 2^5 + 1 \times 2^4 + 1 \times 2^3 + 1 \times 2^2$$
$$+ 0 \times 2^1 + 1 \times 2^0$$

so that each digit doubles in value for every step it moves to the left. In decimal, this number equals 157 – so we have written the same number in two different forms, using two different types of notation.

Binary notation is ideal for electronic systems because it is much easier to distinguish between two possible values of a current, or a voltage, or a magnetic field, than it is to distinguish between more than two. In crude terms, 0 can mean 'no electric current' and 1 can mean 'some electric current', 0 can mean 'no magnetic field' and 1 can mean 'some magnetic field', and so on. In practice engineers set a threshold value, and then 0 means 'below threshold' and 1 means 'above threshold'. By keeping the actual values used for 0 and 1 far enough apart, and setting the threshold in between, there is very little danger of confusing 0 with 1. So devices based on binary notation are robust. That's what makes them digital.

With early computers, the engineers had to struggle to keep the circuit variables within reasonable bounds, and binary made their lives much easier. Modern circuits on silicon chips are precise enough to permit other choices, such as base 3, but the design of digital computers has been based on binary notation for so long now that it generally makes sense to stick to binary, even if alternatives would work. Modern circuits are also very small and very quick. Without some such technological breakthrough in circuit manufacture, the world would have a few thousand computers, rather than billions. Thomas J. Watson, who founded IBM, once said that he didn't think there would be a market for more than about five computers worldwide. At the time he seemed to be talking sense, because in those days the most powerful computers were about the size of a house, consumed as much electricity as a small village, and cost tens of millions of dollars. Only big government organisations, such as the United States Army, could afford one, or make enough use of it. Today a basic, out-of-date mobile phone contains more computing power than anything that was available when Watson made his remark.

The choice of binary representation for digital computers, hence also for digital messages transmitted between computers – and later between almost any two electronic gadgets on the planet – led to the basic unit of information: the *bit*. The name is short for 'binary digit', and one bit of information is one 0 or one 1. It is reasonable to define the information 'contained in' a sequence of binary digits to be the total number of digits in the sequence. So the 8-digit sequence 10011101 contains 8 bits of information.

Shannon realised that simple-minded bit-counting makes sense as a measure of information only if 0s and 1s are like heads and tails with a fair coin, that is, are equally likely to occur. Suppose we know that in some specific circumstances 0 occurs nine times out of ten, and 1 only once. As we read along the string of digits, we expect most digits to be 0. If that expectation is confirmed, we haven't received much information, because this is what we expected anyway. However, if we see 1, that conveys a lot *more* information, because we didn't expect that at all.

We can take advantage of this by encoding the same information more efficiently. If 0 occurs with probability 9/10 and 1 with probability 1/10, we can define a new code like this:

$$000 \to 00 \quad \text{(use whenever possible)}$$
$$00 \to 01 \quad \text{(if no 000 remains)}$$
$$0 \to 10 \quad \text{(if no 00 remains)}$$
$$1 \to 11 \quad \text{(always)}$$

What I mean here is that a message such as

00000000100010000010000001000000000

is first broken up from left to right into blocks that read 000, 00, 0, or 1. With strings of consecutive 0s, we use 000 whenever we can. If not, what's left is either 00 or 0, followed by a 1. So here the message breaks up as

000-000-00-1-000-1-000-00-1-000-000-1-000-000-000

and the coded version becomes

00-00-01-11-00-11-00-01-11-00-00-11-11-00-00-00

The original message has 35 digits, but the encoded version has only 32. The amount of information seems to have decreased.

Sometimes the coded version might be longer: for instance, 111 turns into 111111. But that's rare because 1 occurs only one time in ten on average. There will be quite a lot of 000s, which drop to 00. Any spare 00 changes to 01, the same length; a spare 0 increases the length by changing to 00. The upshot is that in the long run, for randomly chosen messages with the given probabilities of 0 and 1, the coded version is shorter.

My code here is very simple-minded, and a cleverer choice can shorten the message even more. One of the main questions that Shannon wanted to answer was: how efficient can codes of this general type be? If you know the list of symbols that is being used to create a message, and you also know how likely each symbol is, how much can you shorten the message by using a suitable code? His solution was an equation, defining the amount of information in terms of these probabilities.

Suppose for simplicity that the messages use only two symbols 0 and 1, but now these are like flips of a biased coin, so that 0 has probability p of occurring, and 1 has probability $q = 1 - p$. Shannon's analysis led him to a formula for the information content: it should be defined as

$$H = -p \log p - q \log q$$

where log is the logarithm to base 2.

At first sight this doesn't seem terribly intuitive. I'll explain how Shannon derived it in a moment, but the main thing to appreciate at this stage is how H behaves as p varies from 0 to 1, which is shown in Figure 56. The value of H increases smoothly from 0 to 1 as p rises from 0 to $\frac{1}{2}$, and then drops back symmetrically to 0 as p goes from $\frac{1}{2}$ to 1.

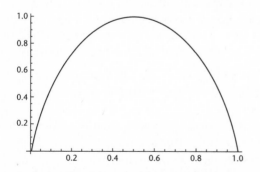

Fig 56 How Shannon information H depends on p. H runs vertically and p runs horizontally.

Shannon pointed out several 'interesting properties' of H, so defined:

- If $p = 0$, in which case only the symbol 1 will occur, the information H is zero. That is, if we are certain which symbol is going to be transmitted to us, receiving it conveys no information whatsoever.

- The same holds when $p=1$. Only the symbol 0 will occur, and again we receive no information.
- The amount of information is largest when $p=q=\frac{1}{2}$, corresponding to the toss of a fair coin. In this case,

$$H = -\tfrac{1}{2}\log\tfrac{1}{2} - \tfrac{1}{2}\log\tfrac{1}{2} = -\log\tfrac{1}{2} = 1$$

bearing in mind that the logarithms are to base 2. That is, one toss of a fair coin conveys one bit of information, as we were originally assuming before we started worrying about coding messages to compress them, and biased coins.

- In all other cases, receiving one symbol conveys *less* information than one bit.
- The more biased the coin becomes, the less information the result of one toss conveys.
- The formula treats the two symbols in exactly the same way. If we exchange p and q, then H stays the same.

All of these properties correspond to our intuitive sense of how much information we receive when we are told the result of a coin toss. That makes the formula a reasonable working definition. Shannon then provided a solid foundation for his definition by listing several basic principles that any measure of information content ought to obey and deriving a unique formula that satisfied them. His set-up was very general: the message could choose from a number of different symbols, occurring with probabilities $p_1, p_2, \ldots, p_n$ where n is the number of symbols. The information H conveyed by a choice of one of these symbols should satisfy:

- H is a continuous function of $p_1, p_2, \ldots, p_n$. That is, small changes in the probabilities should lead to small changes in the amount of information.
- If all of the probabilities are equal, which implies they are all $1/n$, then H should increase if n gets larger. That is, if you are choosing between 3 symbols, all equally likely, then the information you receive should be more than if the choice were between just 2 equally likely symbols; a choice between 4 symbols should convey more information than a choice between 3 symbols, and so on.
- If there is a natural way to break a choice down into two successive

choices, then the original H should be a simple combination of the new Hs.

This final condition is most easily understood using an example, and I've put one in the Notes.[1] Shannon proved that the *only* function H that obeys his three principles is

$$H(p_1, p_2, \ldots, p_n) = -p_1 \log p_1 - p_2 \log p_2 - \ldots - p_n \log p_n$$

or a constant multiple of this expression, which basically just changes the unit of information, like changing from feet to metres.

There is a good reason to take the constant to be 1, and I'll illustrate it in one simple case. Think of the four binary strings 00, 01, 10, 11 as symbols in their own right. If 0 and 1 are equally likely, each string has the same probability, namely $\frac{1}{4}$. The amount of information conveyed by one choice a such a string is therefore

$$H\left(\tfrac{1}{4}, \tfrac{1}{4}, \tfrac{1}{4}, \tfrac{1}{4}\right) = -\tfrac{1}{4}\log\tfrac{1}{4} - \tfrac{1}{4}\log\tfrac{1}{4} - \tfrac{1}{4}\log\tfrac{1}{4} - \tfrac{1}{4}\log\tfrac{1}{4} = -\log\tfrac{1}{4} = 2$$

That is, 2 bits. Which is a sensible number for the information in a length-2 binary string when the choices 0 and 1 are equally likely. In the same way, if the symbols are all length-n binary strings, and we set the constant to 1, then the information content is n bits. Notice that when $n = 2$ we obtain the formula pictured in Figure 56. The proof of Shannon's theorem is too complicated to give here, but it shows that if you accept Shannon's three conditions then there is a single natural way to quantify information.[2] The equation itself is merely a definition: what counts is how it performs in practice.

Shannon used his equation to prove that there is a fundamental limit on how much information a communication channel can convey. Suppose that you are transmitting a digital signal along a phone line, whose capacity to carry a message is at most C bits per second. This capacity is determined by the number of binary digits that the phone line can transmit, and it is not related to the probabilities of various signals. Suppose that the message is being generated from symbols with information content H, also measured in bits per second. Shannon's theorem answers the question: if the channel is noisy, can the signal be encoded so that the proportion of errors is as small as we wish? The answer is that this is always possible, no matter what the noise level is, if H is less than or equal to C. It is not possible if H is greater than C. In fact, the proportion of errors cannot be reduced below the difference $H-C$, no matter which code is employed, but there exist codes that get as close as you wish to that error rate.

Shannon's proof of his theorem demonstrates that codes of the required kind exist, in each of his two cases, but the proof doesn't tell us what those codes are. An entire branch of information science, a mixture of mathematics, computing, and electronic engineering, is devoted to finding efficient codes for specific purposes. It is called coding theory. The methods for coming up with these codes are very diverse, drawing on many areas of mathematics. It is these methods that are incorporated into our electronic gadgetry, be it a smartphone or *Voyager 1*'s transmitter. People carry significant quantities of sophisticated abstract algebra around in their pockets, in the form of software that implements error-correcting codes for mobile phones.

I'll try to convey the flavour of coding theory without getting too tangled in the complexities. One of the most influential concepts in the theory relates codes to multidimensional geometry. It was published by Richard Hamming in 1950 in a famous paper, 'Error detecting and error correcting codes'. In its simplest form, it provides a comparison between strings of binary digits. Consider two such strings, say 10011101 and 10110101. Compare corresponding bits, and count how many times they are different, like this:

$$10\mathbf{01}1101$$
$$10\mathbf{11}0101$$

where I've marked the differences in bold type. Here there are two locations at which the bit-strings differ. We call this number the Hamming distance between the two strings. It can be thought of as the smallest number of one-bit errors that can convert one string into the other. So it is closely related to the likely effect of errors, if these occur at a known average rate. That suggests it might provide some insight into how to detect such errors, and maybe even how to put them right.

Multidimensional geometry comes into play because the strings of a fixed length can be associated with the vertices of a multidimensional 'hypercube'. Riemann taught us how to think of such spaces by thinking of lists of numbers. For example, a space of four dimensions consists of all possible lists of four numbers: (x_1, x_2, x_3, x_4). Each such list is considered to represent a point in the space, and all possible lists can in principle occur.

The separate xs are the coordinates of the point. If the space has 157 dimensions, you have to use lists of 157 numbers: $(x_1, x_2, \ldots, x_{157})$. It is often useful to specify how far apart two such lists are. In 'flat' Euclidean geometry this is done using a simple generalisation of Pythagoras's theorem. Suppose we have a second point $(y_1, y_2, \ldots, y_{157})$ in our 157-dimensional space. Then the distance between the two points is the square root of the sum of the squares of the differences between corresponding coordinates. That is,

$$d = \sqrt{(x_1 - y_1)^2 + (x_2 - y_2)^2 + \ldots + (x_{157} - y_{157})^2}$$

If the space is curved, Riemann's idea of a metric can be used instead.

Hamming's idea is to do something very similar, but the values of the coordinates are restricted to just 0 and 1. Then $(x_1 - y_1)^2$ is 0 if x_1 and y_1 are the same, but 1 if not, and the same goes for $(x_2 - y_2)^2$ and so on. He also omitted the square root, which changes the answer, but in compensation the result is always a whole number, equal to the Hamming distance. This notion has all the properties that make 'distance' useful, such as being zero only when the two strings are identical, and ensuring that the length of any side of a 'triangle' (a set of three strings) is less than or equal to the sum of the lengths of the other two sides.

We can draw pictures of all bit strings of lengths 2, 3, and 4 (and with more effort and less clarity, 5, 6, and possibly even 10, though no one would find that useful). The resulting diagrams are shown in Figure 57.

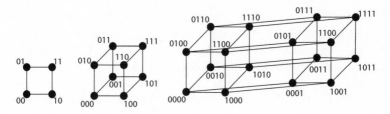

Fig 57 Spaces of all bit-strings of lengths 2, 3, and 4.

The first two are recognisable as a square and a cube (projected on to a plane because it has to be printed on a sheet of paper). The third is a hypercube, the 4-dimensional analogue, and again this has to be projected on to a plane. The straight lines joining the dots have Hamming length 1 – the two strings at either end differ in precisely one location, one coordinate. The Hamming distance between any two strings is the number of such lines in the shortest path that connects them.

Suppose we are thinking of 3-bit strings, living on the corners of a cube. Pick one of the strings, say 101. Suppose the rate of errors is at most one bit in every three. Then this string may either be transmitted unchanged, or it could end up as any of 001, 111, or 100. Each of these differs from the original string in just one location, so its Hamming distance from the original string is 1. In a loose geometrical image, the erroneous strings lie on a 'sphere' centred at the correct string, of radius 1. The sphere consists of just three points, and if we were working in 157-dimensional space with a radius of 5, say, it wouldn't even look terribly spherical. But it plays a similar role to an ordinary sphere: it has a fairly compact shape, and it contains exactly the points whose distance from the centre is less than or equal to the radius.

Suppose we use the spheres to construct a code, so that each sphere corresponds to a new symbol, and that symbol is encoded with the coordinates of the centre of the sphere. Suppose moreover that these spheres don't overlap. For instance, I might introduce a symbol *a* for the sphere centred at 101. This sphere contains four strings: 101, 001, 111, and 100. If I receive any of these four strings, I know that the symbol was originally *a*. At least, that's true provided my other symbols correspond in a similar way to spheres that do not have any points in common with this one.

Now the geometry starts to make itself useful. In the cube, there are eight points (strings) and each sphere contains four of them. If I try to fit spheres into the cube, without them overlapping, the best I can manage is two of them, because $8/4 = 2$. I can actually find another one, namely the sphere centred on 010. This contains 010, 110, 000, 011, none of which are in the first sphere. So I can introduce a second symbol *b* associated with this sphere. My error-correcting code for messages written with *a* and *b* symbols now replaces every *a* by 101, and every *b* by 010. If I receive, say,

101-010-100-101-000

then I can decode the original message as

a-b-a-a-b

despite the errors in the third and fifth string. I just see which of my two spheres the erroneous string belongs to.

All very well, but this multiplies the length of the message by 3, and we already know an easier way to achieve the same result: repeat the message three times. But the same idea takes on a new significance if we work in higher-dimensional spaces. With strings of length 4, the hypercube, there are 16 strings, and each sphere contains 5 points. So it *might* be possible to fit three spheres in without them overlapping. If you try that, it's not actually possible – two fit in but the remaining gap is the wrong shape. But the numbers increasingly work in our favour. The space of strings of length 5 contains 32 strings, and each sphere uses just 6 of them – possibly room for 5, and if not, a better chance of fitting in 4. Length 6 gives us 64 points, and spheres that use 7, so up to 9 spheres might fit in.

From this point on a lot of fiddly detail is needed to work out just what is possible, and it helps to develop more sophisticated methods. But what we are looking at is the analogue, in the space of strings, of the most efficient ways to pack spheres together. And this is a long-standing area of mathematics, about which quite a lot is known. Some of that technique can be transferred from Euclidean geometry to Hamming distances, and when that doesn't work we can invent new methods more suited to the geometry of strings. As an example, Hamming invented a new code, more efficient than any known at the time, which encodes 4-bit strings by converting them into 7-bit strings. It can detect and correct any single-bit error. Modified to an 8-bit code, it can detect, but not correct, any 2-bit error.

This code is called the Hamming code. I won't describe it, but let's do the sums to see if it might be possible. There are 16 strings of length 4, and 128 of length 7. Spheres of radius 1 in the 7-dimensional hypercube contain 8 points. And $128/8 = 16$. So with enough cunning, it might just be possible to squeeze the required 16 spheres into the 7-dimensional hypercube. They would have to fit exactly, because there's no spare room left over. As it happens, such an arrangement exists, and Hamming found it. Without the multidimensional geometry to help, it would be difficult to guess that it existed, let alone find it. Possible, but hard. Even with the geometry, it's not obvious.

Shannon's concept of information provides limits on how efficient codes can be. Coding theory does the other half of the job: finding codes that are as efficient as possible. The most important tools here come from abstract algebra. This is the study of mathematical structures that share the basic arithmetical features of integers or real numbers, but differ from them in significant ways. In arithmetic, we can add numbers, subtract them, and multiply them, to get numbers of the same kind. For the real numbers we can also divide by anything other than zero to get a real number. This is not possible for the integers, because, for example, $\frac{1}{2}$ is not an integer. However, it is possible if we pass to the larger system of rational numbers, fractions. In the familiar number systems, various algebraic laws hold, for example the commutative law of addition, which states that $2+3=3+2$ and the same goes for any two numbers.

The familiar systems share these algebraic properties with less familiar ones. The simplest example uses just two numbers, 0 and 1. Sums and products are defined just as for integers, with one exception: we insist that $1+1=0$, not 2. Despite this modification, all of the usual laws of algebra survive. This system has only two 'elements', two number-like objects. There is exactly one such system whenever the number of elements is a power of any prime number: 2, 3, 4, 5, 7, 8, 9, 11, 13, 16, and so on. Such systems are called Galois fields after the French mathematician Évariste Galois, who classified them around 1830. Because they have finitely many elements, they are suited to digital communications, and powers of 2 are especially convenient because of binary notation.

Galois fields lead to coding systems called *Reed–Solomon codes*, after Irving Reed and Gustave Solomon who invented them in 1960. They are used in consumer electronics, especially CDs and DVDs. They are error-correcting codes based on algebraic properties of polynomials, whose coefficients are taken from a Galois field. The signal being encoded – audio or video – is used to construct a polynomial. If the polynomial has degree n, that is, the highest power occurring is x^n, then the polynomial can be reconstructed from its values at any n points. If we specify the values at more than n points, we can lose or modify some of the values without losing track of which polynomial it is. If the number of errors is not too large, it is still possible to work out which polynomial it is, and decode to get the original data.

In practice the signal is represented as a series of blocks of binary digits. A popular choice uses 255 bytes (8-bit strings) per block. Of these, 223 bytes encode the signal, while the remaining 32 bytes are 'parity symbols', telling us whether various combinations of digits in the uncorrupted data are odd or even. This particular Reed–Solomon code can correct up to 16 errors per block, an error rate just less than 1%.

Whenever you drive along a bumpy road with a CD on the car stereo, you are using abstract algebra, in the form of a Reed–Solomon code, to ensure that the music comes over crisp and clear, instead of being jerky and crackly, perhaps with some parts missing altogether.

Information theory is widely used in cryptography and cryptanalysis – secret codes and methods for breaking them. Shannon himself used it to estimate the amount of coded messages that must be intercepted to stand a chance of breaking the code. Keeping information secret turns out to be more difficult than might be expected, and information theory sheds light on this problem, both from the point of view of the people who want it kept secret and those who want to find out what it is. This issue is important not just to the military, but to everyone who uses the Internet to buy goods or engages in telephone banking.

Information theory now plays a significant role in biology, particularly in the analysis of DNA sequence data. The DNA molecule is a double-helix, formed by two strands that wind round each other. Each strand is a sequence of bases, special molecules that come in four types – adenine, guanine, thymine, and cytosine. So DNA is like a code message written using four possible symbols: A, G, T, and C. The human genome, for example, is 3 billion bases long. Biologists can now find the DNA sequences of innumerable organisms at a rapidly growing rate, leading to a new area of computer science: bioinformatics. This centres on methods for handling biological data efficiently and effectively, and one of its basic tools is information theory.

A more difficult issue is the quality of information, rather than the quantity. The messages 'two plus two make four' and 'two plus two make five' contain exactly the same amount of information, but one is true and the other is false. Paeans of praise for the information age ignore the uncomfortable truth that much of the information rattling around the Internet is misinformation. There are websites run by criminals who want to steal your money, or denialists who want to replace solid science by whichever bee happens to be buzzing around inside their own bonnet.

The vital concept here is not information as such, but meaning. Three billion DNA bases of human DNA information are, literally, meaningless unless we can find out how they affect our bodies and behaviour. On the tenth anniversary of the completion of the Human Genome Project, several leading scientific journals surveyed medical progress resulting so far from listing human DNA bases. The overall tone was muted: a few new cures for diseases have been found so far, but not in the quantity originally predicted. Extracting meaning from DNA information is proving harder than most biologists had hoped. The Human Genome Project was a necessary first step, but it has revealed just how difficult such problems are, rather than solving them.

The notion of information has escaped from electronic engineering and invaded many other areas of science, both as a metaphor and as a technical concept. The formula for information looks very like that for entropy in Boltzmann's approach to thermodynamics; the main differences are logarithms to base 2 instead of natural logarithms, and a change in sign. This similarity can be formalised, and entropy can be interpreted as 'missing information'. So the entropy of a gas increases because we lose track of exactly where its molecules are, and how fast they're moving. The relation between entropy and information has to be set up rather carefully: although the formulas are very similar, the context in which they apply is different. Thermodynamic entropy is a large-scale property of the state of a gas, but information is a property of a signal-producing *source*, not of a signal as such. In 1957 the American physicist Edwin Jaynes, an expert in statistical mechanics, summed up the relationship: thermodynamic entropy can be viewed as an *application* of Shannon information, but entropy itself should not be identified with missing information without specifying the right context. If this distinction is borne in mind, there are valid contexts in which entropy can be viewed as a loss of information. Just as entropy increase places constraints on the efficiency of steam engines, the entropic interpretation of information places constraints on the efficiency of computations. For example, it must take at least 5.8×10^{-23} joules of energy to flip a bit from 0 to 1 or vice versa at the temperature of liquid helium, whatever method you use.

Problems arise when the words 'information' and 'entropy' are used in a more metaphorical sense. Biologists often say that DNA determines 'the information' required to make an organism. There is a sense in which this is almost correct: delete 'the'. However, the metaphorical interpretation of information suggests that once you know the DNA, then you know everything there is to know about the organism. After all, you've got *the* information, right? And for a time many biologists thought that this statement was close to the truth. However, we now know that it is overoptimistic. Even if the information in DNA really did specify the organism uniquely, you would still need to work out how it grows and what the DNA actually does. However, it takes a lot more than a list of DNA codes to create an organism: so-called epigenetic factors must also be taken into account. These include chemical 'switches' that make a segment of DNA code active or inactive, but also entirely different factors that are transmitted from parent to offspring. For human beings, those factors include the culture in which we grow up. So it pays not to be too casual when you use technical terms like 'information'.

16 The imbalance of nature
Chaos Theory

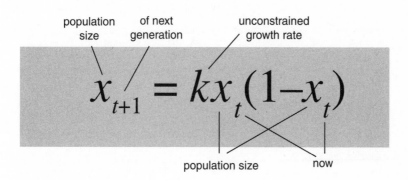

population size → of next generation → unconstrained growth rate

$$x_{t+1} = kx_t(1-x_t)$$

population size — now

What does it say?

It models how a population of living creatures changes from one generation to the next, when there are limits to the available resources.

Why is that important?

It is one of the simplest equations that can generate deterministic chaos – apparently random behaviour with no random cause.

What did it lead to?

The realisation that simple nonlinear equations can create very complex dynamics, and that apparent randomness may conceal hidden order. Popularly known as chaos theory, this discovery has innumerable applications throughout the sciences including the motion of the planets in the Solar System, weather forecasting, population dynamics in ecology, variable stars, earthquake modelling, and efficient trajectories for space probes.

The metaphor of the balance of nature trips readily off the tongue as a description of what the world would do if nasty humans didn't keep interfering. Nature, left to its own devices, would settle down to a state of perfect harmony. Coral reefs would always harbour the same species of colourful fish in similar numbers, rabbits and foxes would learn to share the fields and woodlands so that the foxes would be well fed, most rabbits would survive, and neither population would explode or crash. The world would settle down to a fixed state and stay there. Until the next big meteorite, or a supervolcano, upset the balance.

It's a common metaphor, perilously close to being a cliché. It's also highly misleading. Nature's balance is distinctly wobbly.

We've been here before. When Poincaré was working on King Oscar's prize, the conventional wisdom held that a stable Solar System is one in which the planets follow much the same orbits forever, give or take a harmless bit of jiggling. Technically this is not a steady state, but one in which each planet repeats similar motions over and over again, subject to minor disturbances caused by all the others, but not deviating hugely from what it would have done without them. The dynamics is 'quasiperiodic' – combining several separate periodic motions whose periods are not all multiples of the same time interval. In the realm of planets, that's as close to 'steady' as anyone can hope for.

But the dynamics wasn't like that, as Poincaré belatedly, and to his cost, found out. It could, in the right circumstances, be chaotic. The equations had no explicit random terms, so that in principle the present state completely determined the future state, yet paradoxically the actual motion could appear to be random. In fact, if you asked coarse-grained questions like 'which side of the Sun will it be on?', the answer could be a genuinely random series of observations. Only if you could look infinitely closely would you be able to see that the motion really was completely determined.

This was the first intimation of what we now call 'chaos', which is short for 'deterministic chaos', and quite different from 'random' – even though that's what it can look like. Chaotic dynamics has hidden patterns, but they're subtle; they differ from what we might naturally think of measuring. Only by understanding the causes of chaos can we extract those patterns from an irregular mishmash of data.

As always in science, there were a few isolated precursors, generally viewed as minor curiosities unworthy of serious attention. Only in the 1960s did mathematicians, physicists, and engineers begin to realise just how natural chaos is in dynamics, and how radically it differs from anything envisaged in classical science. We are still learning to appreciate what that tells us, and what to do about it. But already chaotic dynamics, 'chaos theory' in popular parlance, pervades most areas of science. It may even have things to tell us about economics and the social sciences. It's not the answer to everything: only critics ever claimed it was, and that was to make it easier to shoot it down. Chaos has survived all such attacks, and for a good reason: it is absolutely fundamental to all behaviour governed by differential equations, and those are the basic stuff of physical law.

There is chaos in biology, too. One of the first to appreciate that this might be the case was the Australian ecologist Robert May, now Lord May of Oxford and a former president of the Royal Society. He sought to understand how the populations of various species change over time in natural systems such as coral reefs and woodlands. In 1975 May wrote a short article for the journal *Nature*, pointing out that the equations typically used to model changes to animal and plant populations could produce chaos. May didn't claim that the models he was discussing were accurate representations of what real populations did. His point was more general: chaos was natural in models of that kind, and this had to be borne in mind.

The most important consequence of chaos is that irregular behaviour need not have irregular causes. Previously, if ecologists noticed that some population of animals was fluctuating wildly, they would look for some external cause – also presumed to be fluctuating wildly, and generally labelled 'random'. The weather, perhaps, or a sudden influx of predators from elsewhere. May's examples showed that the internal workings of the animal populations could generate irregularity without outside help.

His main example was the equation that decorates the opening of this chapter. It is called the logistic equation, and it is a simple model of a population of animals in which the size of each generation is determined by the previous one. 'Discrete' means that the flow of time is counted in generations, and is thus an integer. So the model is similar to a differential equation, in which time is a continuous variable, but conceptually and computationally simpler. The population is measured as a fraction of some overall large value, and can therefore be represented by a real number that lies between 0 (extinction) and 1 (the theoretical maximum that the system can sustain). Letting time t tick in integer steps, corresponding to generations, this number is x_t in generation t. The logistic equation states that

$$x_{t+1} = kx_t(1 - x_t)$$

where k is a constant. We can interpret k as the growth rate of the population when diminishing resources do not slow it down.[1]

We start the model at time 0 with an initial population x_0. Then we use the equation with $t = 0$ to calculate x_1, then we set $t = 1$ and compute x_2, and so on. Without even doing the sums we can see straight away that, for any fixed growth rate k, the population size of generation zero completely determines the sizes of all succeeding generations. So the model is *deterministic*: knowledge of the present determines the future uniquely and exactly.

So what *is* the future? The 'balance of nature' metaphor suggests that the population should settle to a steady state. We can even calculate what that steady state should be: just set the population at time $t + 1$ to be the same as that at time t. This leads to two steady states: populations 0 and 1-1/k. A population of size 0 is extinct, so the other value should apply to an existing population. Unfortunately, although this is a steady state, it can be unstable. If it is, then in practice you'll never see it: it's like trying to balance a pencil vertically on its sharpened point. The slightest disturbance will cause it to topple. The calculations show that the steady state is unstable when k is bigger than 3.

What, then, *do* we see in practice? Figure 58 shows a typical 'time series' for the population when $k = 4$. It's not steady: it's all over the place. However, if you look closely there are hints that the dynamics is not completely random. Whenever the population gets really big, it immediately crashes to a very low value, and then grows in a regular manner (roughly exponentially) for the next two or three generations: see the short arrows in Figure 58. And something interesting happens whenever the population gets close to 0.75 or thereabouts: it oscillates alternately above and below that value, and the oscillations grow giving a characteristic zigzag shape, getting wider towards the right: see the longer arrows in the figure.

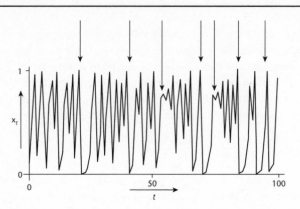

Fig 58 Chaotic oscillations in a model animal population. Short arrows show crashes followed by short-term exponential growth. Longer arrows show unstable oscillations.

Despite these patterns, there is a sense in which the behaviour is truly random – but only when you throw away some of the detail. Suppose we assign the symbol H (heads) whenever the population is bigger than 0.5, and T (tails) when it's less than 0.5. This particular set of data begins with the sequence THTHTHHTHHTTHH and continues unpredictably, just like a random sequence of coin tosses. This way of coarsening the data, by looking at specific ranges of values and noting only which range the population belongs to, is called symbolic dynamics. In this case, it is possible to prove that, for almost all initial population values x_0, the sequence of heads and tails is in all respects like a typical sequence of random tosses of a fair coin. Only when we look at the exact values do we start to see some patterns.

It's a remarkable discovery. A dynamical system can be completely deterministic, with visible patterns in detailed data, yet a coarse-grained view of the same data can be random – in a provable, rigorous sense. Determinism and randomness are not opposites. In some circumstances, they can be entirely compatible.

May didn't invent the logistic equation, and he didn't discover its astonishing properties. He didn't claim to have done either of those things. His aim was to alert workers in the life sciences, especially ecologists, to the remarkable discoveries emerging in the physical sciences and mathematics: discoveries that fundamentally change the way scientists should think about observational data. We humans may have trouble solving equations based on simple rules, but nature doesn't have to solve the equations the way we do. It just obeys the rules. So it can do things that strike us as being complicated, for simple reasons.

Chaos emerged from a topological approach to dynamics, orchestrated in particular by the American mathematician Stephen Smale and the Russian mathematician Vladimir Arnold in the 1960s. Both were trying to find out what types of behaviour were typical in differential equations. Smale was motivated by Poincaré's strange results on the three-body problem (Chapter 4), and Arnold was inspired by related discoveries of his former research supervisor Andrei Kolmogorov. Both quickly realised why chaos is common: it is a natural consequence of the geometry of differential equations, as we'll see in a moment.

As interest in chaos spread, examples were spotted lurking unnoticed in earlier scientific papers. Previously considered to be just isolated weird effects, these examples now slotted into a broader theory. In the 1940s the English mathematicians John Littlewood and Mary Cartwright had seen traces of chaos in electronic oscillators. In 1958 Tsuneji Rikitake of Tokyo's Association for the Development of Earthquake Prediction had found chaotic behaviour in a dynamo model of the Earth's magnetic field. And in 1963 the American meteorologist Edward Lorenz had pinned down the nature of chaotic dynamics in considerable detail, in a simple model of atmospheric convection motivated by weather-forecasting. These and other pioneers had pointed the way; now all of their disparate discoveries were starting to fit together.

In particular, the circumstances that led to chaos, rather than something simpler, turned out to be geometric rather than algebraic. In the logistic model with $k = 4$, both extremes of the population, 0 and 1, move to 0 in the next generation, while the midpoint, $\frac{1}{2}$, moves to 1. So at each time-step the interval from 0 to 1 is stretched to twice its length, folded in half, and slapped down in its original location. This is what a cook does to dough when making bread, and by thinking about dough being kneaded, we gain a handle on chaos. Imagine a tiny speck in the logistic dough – a raisin, say. Suppose that it happens to lie on a periodic cycle, so that after a certain number of stretch-and-fold operations it returns to where it started. Now we can see why this point is unstable. Imagine another raisin, initially very close to the first one. Each stretch moves it further away. For a time, though, it doesn't move far enough away to stop tracking the first raisin. When the dough is folded, both raisins end up in the same layer. So next time, the second raisin has moved even further away from the first. This is why the periodic state is unstable:

stretching moves all nearby points *away* from it, not towards it. Eventually the expansion becomes so great that the two raisins end up in different layers when the dough is folded. After that, their fates are pretty much independent of each other. Why does a cook knead dough? To mix up the ingredients (including trapped air). If you mix stuff up, the individual particles have to move in a very irregular way. Particles that start close together end up far apart; points far apart may be folded back to be close together. In short, chaos is the natural result of *mixing*.

I said at the start of this chapter that you don't have anything chaotic in your kitchen, except perhaps that dishwasher. I lied. You probably have several chaotic gadgets: a food processor, an egg-beater. The blade of the food processor follows a very simple rule: go round and round, fast. The food interacts with the blade: it ought to do something simple too. But it doesn't go round and round: it gets mixed up. As the blade cuts through the food, some bits go one side of it, some go the other side: locally, the food gets pulled apart. But it doesn't escape from the mixing bowl, so it all gets folded back in on itself.

Smale and Arnold realised that all chaotic dynamics is like this. They didn't phrase their results in quite that language, mind you: 'pulled apart' was 'positive Liapunov exponent' and 'folded back' was 'the system has a compact domain'. But in fancy language, they were saying that chaos is like mixing dough.

This also explains something else, noticed especially by Lorenz in 1963. Chaotic dynamics is sensitive to initial conditions. However close the two raisins are to begin with, they eventually get pulled so far apart that their subsequent movements are independent. This phenomenon is often called the butterfly effect: a butterfly flaps its wings, and a month later the weather is completely different from what it would otherwise have been. The phrase is generally credited to Lorenz. He didn't introduce it, but something similar featured in the title of one of his lectures. However, someone else invented the title for him, and the lecture wasn't about the famous 1963 article, but a lesser-known one from the same year.

Whatever the phenomenon is called, it has an important practical consequence. Although chaotic dynamics is in principle deterministic, in practice it becomes unpredictable very quickly, because any uncertainty in the exact initial state grows exponentially fast. There is a prediction horizon beyond which the future cannot be foreseen. For weather, a familiar system whose standard computer models are known to be chaotic, this horizon is a few days ahead. For the Solar System, it is tens of millions of years ahead. For simple laboratory toys, such as a double pendulum (a pendulum hung from the bottom of another one) it is a few seconds ahead. The long-held assumption that 'deterministic' and 'predictable' are the same is wrong. It would be valid if the present state of a system could be measured with perfect accuracy, but that's not possible.

The short-term predictability of chaos can be used to distinguish it from pure randomness. Many different techniques have been devised to make this distinction, and to work out the underlying dynamics if the system is behaving deterministically but chaotically.

Chaos now has applications in every branch of science, from astronomy to zoology. In Chapter 4 we saw how it is leading to new, more efficient trajectories for space missions. In broader terms, astronomers Jack Wisdom and Jacques Laskar have shown that the dynamics of the Solar System is chaotic. If you want to know whereabouts in its orbit Pluto will be in 10,000,000 AD – forget it. They have also shown that the Moon's tides stabilise the Earth against influences that would otherwise lead to chaotic motion, causing rapid shifts of climate from warm periods to ice ages and back again. So chaos theory demonstrates that, without the Moon, the Earth would be a pretty unpleasant place to live. This feature of our planetary neighbourhood is often used to argue that the evolution of life on a planet requires a stabilising Moon, but this is an overstatement. Life in the oceans would scarcely notice if the planet's axis changed over a period of millions of years. Life on land would have plenty of time to migrate elsewhere, unless it got trapped somewhere that lacked a land route to a place where conditions were more suitable. Climate change is happening much faster right now than anything that a change in axial tilt could cause.

May's suggestion that irregular population dynamics in an ecosystem might sometimes be caused by internal chaos, rather than extraneous randomness, has been verified in laboratory versions of several real-world ecosystems. In 1995 a team headed by American ecologist James Cushing found chaotic dynamics in populations of the flour beetle (or bran bug) *Tribolium castaneum*, which can infest stores of flour.[2] In 1999, Dutch biologists Jef Huisman and Franz Weissing applied chaos to the 'paradox of the plankton', the unexpected diversity of plankton species.[3] A standard principle in ecology, the principle of competitive exclusion, states that an ecosystem cannot contain more species than the number of environmental niches, ways to make a living. Plankton appear to violate this principle: the number of niches is small, but the number of species is in the thousands.

They traced this to a loophole in the derivation of the principle of competitive exclusion: the assumption that populations are steady. If the populations can change over time, then the mathematical derivation from the usual model fails, and intuitively different species can occupy the same niche by taking turns – not by conscious cooperation, but by one species temporarily taking over from another and undergoing a population boom, while the displaced species drops to a small population, Figure 59.

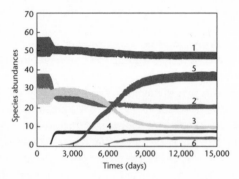

Fig 59 Six species sharing three resources. The bands are closely spaced chaotic oscillations. Courtesy of Jef Huisman and Franz Weissing.

In 2008, Huisman's team published the results of a laboratory experiment with a miniature ecology based on one found in the Baltic Sea, involving bacteria and several kinds of plankton. A six-year study revealed chaotic dynamics in which populations fluctuated wildly, often becoming 100 times as large for a time and then crashing. The usual methods for detecting chaos confirmed its presence. There was even a butterfly effect: the system's prediction horizon was a few weeks.[4]

There are applications of chaos that impinge on everyday life, but they mostly occur in manufacturing processes and public services, rather than being incorporated into gadgets. The discovery of the butterfly effect has changed the way weather forecasts are carried out. Instead of putting all of the computational effort into refining a single prediction, meteorologists now run many forecasts, making different tiny random changes to the observations provided by weather balloons and satellites before starting each run. If all of these forecasts agree, then the prediction is likely to be accurate; if they differ significantly, the weather is in a less predictable state. The forecasts themselves have been improved by several other advances, notably in calculating the influence of the oceans on the state of the atmosphere, but the main role of chaos has been to warn forecasters not to expect too much and to quantify how likely a forecast is to be correct.

Industrial applications include a better understanding of mixing processes, which are widely used to make medicinal pills or mix food ingredients. The active medicine in a pill usually occurs in very small quantities, and it has to be mixed with some inert substance. It's important to get enough of the active ingredient in each pill, but not too much. A mixing machine is like a giant food processor, and like the food processor, its dynamics is deterministic but chaotic. The mathematics of chaos has provided a new understanding of mixing processes and led to some improved designs. The methods used to detect chaos in data have inspired new test equipment for the wire used to make springs, improving efficiency in spring- and wire-making. The humble spring has many vital uses: it can be found in mattresses, cars, DVD players, even ballpoint pens. Chaotic control, a technique that uses the butterfly effect to keep dynamic behaviour stable, is showing promise in the design of more efficient and less intrusive heart pacemakers.

Overall, though, the main impact of chaos has been on scientific thinking. In the forty years or so since its existence started to be widely appreciated, chaos has changed from a minor mathematical curiosity into a basic feature of science. We can now study many of nature's irregularities without resorting to statistics, by teasing out the hidden patterns that characterise deterministic chaos. This is just one of the ways in which modern dynamical systems theory, with its emphasis on nonlinear behaviour, is causing a quiet revolution in the way scientists think about the world.

17 The Midas formula
Black–Scholes Equation

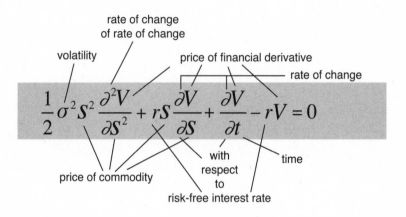

$$\frac{1}{2}\sigma^2 S^2 \frac{\partial^2 V}{\partial S^2} + rS \frac{\partial V}{\partial S} + \frac{\partial V}{\partial t} - rV = 0$$

volatility — rate of change of rate of change — price of financial derivative — rate of change — price of commodity — with respect to — risk-free interest rate — time

What does it say?

It describes how the price of a financial derivative changes over time, based on the principle that when the price is correct, the derivative carries no risk and no one can make a profit by selling it at a different price.

Why is that important?

It makes it possible to trade a derivative before it matures by assigning an agreed 'rational' value to it, so that it can become a virtual commodity in its own right.

What did it lead to?

Massive growth of the financial sector, ever more complex financial instruments, surges in economic prosperity punctuated by crashes, the turbulent stock markets of the 1990s, the 2008–9 financial crisis, and the ongoing economic slump.

Since the turn of the century the greatest source of growth in the financial sector has been in financial instruments known as derivatives. Derivatives are not money, nor are they investments in stocks and shares. They are investments in investments, promises about promises. Derivatives traders use virtual money, numbers in a computer. They borrow it from investors who have probably borrowed it from somewhere else. Often they haven't borrowed it at all, not even virtually: they have clicked a mouse to agree that they *will* borrow the money if it ever becomes necessary. But they have no intention of letting it become necessary; they will sell the derivative before that happens. The lender – hypothetical lender, since the loan will never occur, for the same reason – probably doesn't actually have the money either. This is finance in cloud cuckoo land, yet it has become the standard practice of the world's banking system.

Unfortunately, the consequences of derivatives trading do, ultimately, turn into real money, and real people suffer. The trick works, most of the time, because the disconnect with reality has no notable effect, other than making a few bankers and traders extremely rich as they siphon off real money from the virtual pool. Until things go wrong. Then the pigeons come home to roost, bearing with them virtual debts that have to be paid with real money. By everyone else, naturally.

This is what triggered the banking crisis of 2008–9, from which the world's economies are still reeling. Low interest rates and enormous personal bonus payments encouraged bankers and their banks to bet ever larger sums of virtual money on ever more complex derivatives, ultimately secured – so they believed – in the property market, houses and businesses. As the supply of suitable property and people to buy it began to dry up, the financial world's leaders needed to find new ways to convince shareholders that they were creating profit, in order to justify and finance their bonuses. So they started trading packages of debt, also allegedly secured, somewhere down the line, on real property. Keeping the scheme going demanded the continued purchase of property, to increase the pool of collateral. So the banks started selling mortgages to people whose ability to repay them was increasingly doubtful. This was the subprime mortgage market, 'subprime' being a euphemism for 'likely to default'. Which soon became 'certain to default'.

The banks behaved like one of those cartoon characters who wanders off the edge of a cliff, hovers in space until he looks down, and only then plunges to the ground. It all seemed to be going nicely until the bankers asked themselves whether multiple accounting with non-existent money and overvalued assets was sustainable, wondered what the real value of their holdings in derivatives was, and realised that they didn't have a clue. Except that it was definitely a lot less than they'd told shareholders and government regulators.

As the dreadful truth dawned, confidence plummeted. This depressed the housing market, so the assets against which the debts were secured started to lose their value. At this point the whole system became trapped in a positive feedback loop, in which each downward revision of value caused it to be revised even further downward. The end result was the loss of about 17 trillion dollars. Faced with the prospect of the total collapse of the world financial system, trashing depositors' savings and making the Great Depression of 1929 look like a garden party, governments were forced to bail out the banks, which were on the verge of bankruptcy. One, Lehman Brothers, was allowed to go under, but the loss of confidence was so great that it seemed unwise to repeat the lesson. So taxpayers stumped up the money, and a lot of it was real money. The banks grabbed the cash with both hands, and then tried to pretend that the catastrophe hadn't been their fault. They blamed government regulators, despite having campaigned against regulation: an interesting case of 'It's your fault: you let us do it.'

How did the biggest financial train wreck in human history come about?

Arguably, one contributor was a mathematical equation.

The simplest derivatives have been around for a long time. They are known as futures and options, and they go back to the eighteenth century at the Dojima rice exchange in Osaka, Japan. The exchange was founded in 1697, a time of great economic prosperity in Japan, when the upper classes, the samurai, were paid in rice, not money. Naturally there emerged a class of ricebrokers who traded rice as though it were money. As the Osaka merchants strengthened their grip on rice, the country's staple food, their activities had a knock-on effect on the commodity's price. At the same time, the financial system was beginning to shift to hard cash, and the combination proved deadly. In 1730 the price of rice dropped through the floor.

Ironically, the trigger was poor harvests. The samurai, still wedded to payment in rice, but watchful of the growth of money, started to panic. Their favoured 'currency' was rapidly losing its value. Merchants exacerbated the problem by artificially keeping rice out of the market, squirrelling away huge quantities in warehouses. Although it might seem that this would increase the monetary price of rice, it had the opposite effect, because the samurai were treating rice as a currency. They could not eat anything remotely approaching the amount of rice they owned. So while ordinary people starved, the merchants stockpiled rice. Rice became so scarce that paper money took over, and it quickly became more desirable than rice because it was possible actually to lay hands on it. Soon the Dojima merchants were running what amounted to a gigantic banking system, holding accounts for the wealthy and determining the exchange rate between rice and paper money.

Eventually the government realised that this arrangement handed far too much power to the rice merchants, and reorganised the Rice Exchange along with most other parts of the country's economy. In 1939 the Rice Exchange was replaced by the Government Rice Agency. But while the Rice Exchange existed, the merchants invented a new kind of contract to even out the large swings in the price of rice. The signatories guaranteed to buy (or sell) a specified quantity of rice at a specified future date for a specified price. Today these instruments are known as futures or options. Suppose a merchant agrees to buy rice in six months' time at an agreed price. If the market price has risen above the agreed one by the time the option falls due, he gets the rice cheap and immediately sells it at a profit. On the other hand, if the price is lower, he is committed to buying rice at a higher price than its market value and makes a loss.

Farmers find such instruments useful because they actually want to sell a real commodity: rice. People using rice for food, or manufacturing foodstuffs that use it, want to buy the commodity. In this sort of transaction, the contract reduces the risk to both parties – though at a price. It amounts to a form of insurance: a guaranteed market at a guaranteed price, independent of shifts in the market value. It's worth paying a small premium to avoid uncertainty. But most investors took out contracts in rice futures with the sole aim of making money, and the last thing the investor wanted was tons and tons of rice. They always sold it before they had to take delivery. So the main role of futures was to fuel financial speculation, and this was made worse by the use of rice as currency. Just as today's gold standard creates artificially high prices for a substance (gold) that has little intrinsic value, and thereby fuels demand for it, so the price of rice became governed by the trading of futures rather than the trading of rice itself. The contracts were a form of gambling, and soon the contracts themselves acquired a value, and could be traded as though they were real commodities. Moreover, although the amount of rice was limited by what the farmers could grow, there was no limit to the number of contracts for rice that could be issued.

The world's major stock markets were quick to spot an opportunity to convert smoke and mirrors into hard cash, and they have traded futures ever since. At first, this practice did not of itself cause enormous economic problems, although it sometimes led to instability rather than the stability that is often asserted to justify the system. But around the year 2000, the world's financial sector began to invent ever more elaborate variants on the futures theme, complex 'derivatives' whose value was based on hypothetical future movements of some asset. Unlike futures, for which the asset, at least, was real, derivatives might be based on an asset that was itself a derivative. No longer were banks buying and selling bets on the future price of a commodity like rice; they were buying and selling bets on the future price of a *bet*.

It quickly became big business. In 1998 the international financial system traded roughly $100 trillion in derivatives. By 2007 this had grown to one quadrillion US dollars. Trillions, quadrillions... we know these are large numbers, but how large? To put this figure in context, the total value of all the products made by the world's manufacturing industries, for the last thousand years, is about 100 trillion US dollars, adjusted for inflation. That's one tenth of one year's derivatives trading. Admittedly the bulk of industrial production has occurred in the past fifty years, but even so, this is a staggering amount. It means, in particular, that the derivatives trades consist almost entirely of money that does not actually exist – virtual money, numbers in a computer, with no link to anything in the real world. In fact, these trades *have* to be virtual: the total amount of money in circulation, worldwide, is completely inadequate to pay the amounts that are being traded at the click of a mouse. By people who have no interest in the commodity concerned, and wouldn't know what to do with it if they took delivery, using money that they don't actually possess.

You don't need to be a rocket scientist to suspect that this is a recipe for disaster. Yet for a decade, the world economy grew relentlessly on the back of derivatives trading. Not only could you get a mortgage to buy a house: you could get more than the house was worth. The bank didn't even bother to check what your true income was, or what other debts you had. You could get a 125% self-certified mortgage – meaning you told the bank what you could afford and it didn't ask awkward questions – and spend the surplus on a holiday, a car, plastic surgery, or crates of beer. Banks went out of their way to persuade customers to take out loans, even when they didn't need them.

What they thought would save them if a borrower defaulted on their repayments was straightforward. Those loans were secured on your house. House prices were soaring, so that missing 25% of equity would soon become real; if you defaulted, the bank could seize your house, sell it, and get its loan back. It seemed foolproof. Of course it wasn't. The bankers didn't ask themselves what would happen to the price of housing if hundreds of banks were all trying to sell millions of houses at the same time. Nor did they ask whether prices could continue to rise significantly faster than inflation. They genuinely seemed to think that house prices could rise 10–15% in real terms every year, indefinitely. They were still urging regulators to relax the rules and allow them to lend even more money when the bottom dropped out of the property market.

Many of today's most sophisticated mathematical models of financial systems can be traced back to Brownian motion, mentioned in Chapter 12. When viewed through a microscope, small particles suspended in a fluid jiggle around erratically, and Einstein and Smoluchowski developed mathematical models of this process and used them to establish the existence of atoms. The usual model assumes that the particle receives random kicks through distances whose probability distribution is normal, a bell curve. The direction of each kick is uniformly distributed – any direction has the same chance of happening. This process is called a random walk. The model of Brownian motion is a continuum version of such random walks, in which the sizes of the kicks and the time between successive kicks become arbitrarily small. Intuitively, we consider infinitely many infinitesimal kicks.

The statistical properties of Brownian motion, over large numbers of trials, are determined by a probability distribution, which gives the likelihood that the particle ends up at a particular location after a given time. This distribution is radially symmetric: the probability depends only on how far the point is from the origin. Initially the particle is very likely to be close to the origin, but as time passes, the range of likely positions spreads out as the particle gets more chance to explore distant regions of space. Remarkably, the time evolution of this probability distribution obeys the heat equation, which in this context is often called the diffusion equation. So the probability spreads just like heat.

After Einstein and Smoluchowski published their work, it turned out that much of the mathematical content had been derived earlier, in 1900, by the French mathematician Louis Bachelier in his PhD thesis. But Bachelier had a different application in mind: the stock and option markets. The title of his thesis was *Théorie de la speculation* ('Theory of Speculation'). The work was not received with wild praise, probably because its subject-matter was far outside the normal range of mathematics at that period. Bachelier's supervisor was the renowned and formidable mathematician Henri Poincaré, who declared the work to be 'very original'. He also gave the game way somewhat, by adding, with reference to the part of the thesis that derived the normal distribution for errors: 'It is regrettable that M. Bachelier did not develop this part of his thesis further.' Which any mathematician would interpret as 'that was the place where the mathematics started to get really interesting, and if only he'd done more work on that, rather than on fuzzy ideas about the stock market, it would have been easy to give him a much better grade.' The thesis was graded 'honorable', a pass; it was even published. But it did not get the top grade of 'très honorable'.

Bachelier in effect pinned down the principle that fluctuations of the stock market follow a random walk. The sizes of successive fluctuations conform to a bell curve, and the mean and standard deviation can be estimated from market data. One implication is that large fluctuations are very improbable. The reason is that the tails of the normal distribution die down very fast indeed: faster than exponential. The bell curve decreases towards zero at a rate that is exponential in the *square* of x. Statisticians (and physicists and market analysts) talk of two-sigma fluctuations, three-sigma ones, and so on. Here sigma (σ) is the standard deviation, a measure of how wide the bell curve is. A three-sigma fluctuation, say, is one that deviates from the mean by at least three times the standard deviation. The mathematics of the bell curve lets us assign probabilities to these 'extreme events', see Table 3.

minimum size of fluctuation	probability
σ	0.3174
2σ	0.0456
3σ	0.0027
4σ	0.000063
5σ	0.0000006

Table 3   Probabilities of many-sigma events.

The upshot of Bachelier's Brownian motion model is that large stock market fluctuations are so rare that in practice they should never happen. Table 3 shows that a five-sigma event, for example, is expected to occur about six times in every 10 million trials. However, stock market data show that they are far more common than that. Stock in Cisco Systems, a world leader in communications, has undergone ten 5-sigma events in the last twenty years, whereas Brownian motion predicts 0.003 of them. I picked this company at random and it's in no way unusual. On Black Monday (19 October 1987) the world's stock markets lost more than 20% of their value within a few hours; an event this extreme should have been virtually impossible.

The data suggest unequivocally that extreme events are nowhere near as rare as Brownian motion predicts. The probability distribution does not die way exponentially (or faster); it dies away like a power-law curve x^{-a} for some positive constant a. In the financial jargon, such a distribution is said to have a *fat tail*. Fat tails indicate increased levels of risk. If your investment has a five-sigma expected return, then assuming Brownian motion, the chance that it will fail is less than one in a million. But if tails are fat, it might be much larger, maybe one in a hundred. That makes it a much poorer bet.

A related term, made popular by Nassim Nicholas Taleb, an expert in mathematical finance, is 'black swan event'. His 2007 book *The Black Swan* became a major bestseller. In ancient times, all known swans were white. The poet Juvenal refers to something as 'a rare bird in the lands, and very like a black swan', and he meant that it was impossible. The phrase was widely used in the sixteenth century, much as we might refer to a flying pig. But in 1697, when the Dutch explorer Willem de Vlamingh went to the aptly named Swan River in Western Australia, he found masses of black swans. The phrase changed its meaning, and now refers to an assumption that appears to be grounded in fact, but might at any moment turn out to be wildly mistaken. Yet another term current is X-event, 'extreme event'.

These early analyses of markets in mathematical terms encouraged the seductive idea that the market could be modelled mathematically, creating a rational and safe way to make unlimited sums of money. In 1973 it seemed that the dream might become real, when Fischer Black and Myron Scholes introduced a method for pricing options: the Black–Scholes equation. Robert Merton provided a mathematical analysis of their model in the same year, and extended it. The equation is:

$$\tfrac{1}{2}(\sigma S)^2 \frac{\partial^2 V}{\partial S^2} + rS\frac{\partial V}{\partial S} + \frac{\partial V}{\partial t} - rV = 0$$

It involves five distinct quantities – time t, the price S of the commodity, the price V of the derivative, which depends on S and t, the risk-free interest rate r (the theoretical interest that can be earned by an investment with zero risk, such as government bonds), and the volatility σ^2 of the stock. It is also mathematically sophisticated: a second-order partial differential equation like the wave and heat equations. It expresses the rate of change of the price of the derivative, with respect to time, as a linear combination of three terms: the price of the derivative itself, how fast that changes relative to the stock price, and how that change accelerates. The other variables appear in the coefficients of those terms. If the terms representing the price of the derivative and its rate of change were omitted, the equation would be exactly the heat equation, describing how the price of the option diffuses through stock-price-space. This traces back to Bachelier's assumption of Brownian motion. The other terms take additional factors into account.

The Black–Scholes equation was derived as a consequence of a number of simplifying financial assumptions – for instance, that there are no transaction costs and no limits on short-selling, and that it is possible to lend and borrow money at a known, fixed, risk-free interest rate. The approach is called arbitrage pricing theory, and its mathematical core goes back to Bachelier. It assumes that market prices behave statistically like Brownian motion, in which both the rate of drift and the market volatility are constant. Drift is the movement of the mean, and volatility is financial jargon for standard deviation, a measure of average divergence from the mean. This assumption is so common in the financial literature that it has become an industry standard.

There are two main kinds of option. In a put option, the buyer of the option purchases the right to sell a commodity or financial instrument at a specified time for an agreed price, if they so wish. A call option is similar, but it confers the right to buy instead of sell. The Black–Scholes equation has explicit solutions: one formula for put options, another for call options.[1] If such formulas had not existed, the equation could still have been solved numerically and implemented as software. However, the formulas make it straightforward to calculate the recommended price, as well as yielding important theoretical insights.

The Black–Scholes equation was devised to bring a degree of rationality to the futures market, which it does very effectively under normal market conditions. It provides a systematic way to calculate the value of an option *before it matures*. Then it can be sold. Suppose, for instance, that a merchant contracts to buy 1000 tons of rice in 12 months' time at a price of 500 per ton – a call option. After five months she decides to sell the option to anyone willing to buy it. Everyone knows how the market price for rice has been changing, so how much is that contract worth right now? If you start trading such options without knowing the answer, you're in trouble. If the trade loses money, you're open to the accusation that you got the price wrong and your job could be at risk. So what should the price be? Trading by the seat of your pants ceases to be an option when the sums involved are in the billions. There has to be an agreed way to price an option at any time before maturity. The equation does just that. It provides a formula, which anyone can use, and if your boss uses the same formula, he will get the same result that you did, provided you didn't make errors of arithmetic. In practice, both of you would use a standard computer package.

The equation was so effective that it won Merton and Scholes the 1997 Nobel Prize in Economics.[2] Black had died by then, and the rules of the prize prohibit posthumous awards, but his contribution was explicitly cited by the Swedish Academy. The effectiveness of the equation depended on the market behaving itself. If the assumptions behind the model ceased to hold, it was no longer wise to use it. But as time passed and confidence grew, many bankers and traders forgot that; they used the equation as a kind of talisman, a bit of mathematical magic that protected them against criticism. Black–Scholes not only provides a price that is reasonable under normal conditions; it also covers your back if the trade goes belly-up. Don't blame me, boss, I used the industry standard formula.

The financial sector was quick to see the advantages of the Black–Scholes equation and its solutions, and equally quick to develop a host of related equations with different assumptions aimed at different financial instruments. The then-sedate world of conventional banking could use the equations to justify loans and trades, always keeping an eye open for potential trouble. But less conventional businesses would soon follow, and they had the faith of a true convert. To them, the possibility of the model going wrong was inconceivable. It became known as the Midas formula – a recipe for making everything turn to gold. But the financial sector forgot how the story of King Midas ended.

The darling of the financial sector, for several years, was a company called Long Term Capital Management (LTCM). It was a hedge fund, a private fund that spreads its investments in a way that is intended to protect investors when the market goes down, and make big profits when it goes up. It specialised in trading strategies based on mathematical models, including the Black–Scholes equation and its extensions, together with techniques such as arbitrage, which exploits discrepancies between the prices of bonds and the value that can actually be realised. Initially LTCM was a spectacular success, yielding returns in the region of 40% per year until 1998. At that point it lost $4.6 billion in under four months, and the Federal Reserve Bank persuaded its major creditors to bail it out to the tune of $3.6 billion. Eventually the banks involved got their money back, but LTCM was wound up in 2000.

What went wrong? There are as many theories as there are financial commentators, but the consensus is that the proximate cause of LTCM's failure was the Russian financial crisis of 1998. Western markets had invested heavily in Russia, whose economy was heavily dependent on oil exports. The Asian financial crisis of 1997 caused the price of oil to slump, and the main casualty was the Russian economy. The World Bank provided a loan of $22.6 billion to prop the Russians up.

The ultimate cause of LTCM's demise was already in place on the day it started trading. As soon as reality ceased to obey the assumptions of the model, LTCM was in deep trouble. The Russian financial crisis threw a spanner in the works that demolished almost all of those assumptions. Some factors had a bigger effect than others. Increased volatility was one of them. Another was the assumption that extreme fluctuations hardly ever occur: no fat tails. But the crisis sent the markets into turmoil, and in the panic, prices dropped by huge amounts – many sigmas – in seconds. Because all of the factors concerned were interrelated, these events triggered other rapid changes, so rapid that traders could not possibly know the state of the market at any instant. Even if they wanted to behave rationally, which people don't do in a general panic, they had no basis upon which to do so.

If the Brownian model is right, events as extreme as the Russian financial crisis should occur no more often than once a century. I can remember seven from personal experience in the past 40 years: over-investment in property, the former Soviet Union, Brazil, property (again), property (yet again), dotcom companies, and... oh, yes, property.

With hindsight, the collapse of LTCM was a warning. The dangers of trading by formula in a world that did not obey the cosy assumptions behind the formula were duly noted – and quickly ignored. Hindsight is all very well, but anyone can see the danger after a crisis has struck. What about foresight? The orthodox claim about the recent global financial crisis is that, like the first swan with black feathers, no one saw it coming.

That's not entirely true.

The International Congress of Mathematicians is the largest mathematical conference in the world, taking place every four years. In August 2002 it took place in Beijing, and Mary Poovey, professor of humanities and director of the Institute for the Production of Knowledge at New York University, gave a lecture with the title 'Can numbers ensure honesty?'[3] The subtitle was 'unrealistic expectations and the US accounting scandal', and it described the recent emergence of a 'new axis of power' in world affairs:

> This axis runs through large multinational corporations, many of which avoid national taxes by incorporating in tax havens like Hong Kong. It runs through investment banks, through nongovernmental organizations like the International Monetary Fund, through state and corporate pension funds, and through the wallets of ordinary investors. This axis of financial power contributes to economic catastrophes like the 1998 meltdown in Japan and Argentina's default in 2001, and it leaves its traces in the daily gyrations of stock indexes like the Dow Jones Industrials and London's Financial Times Stock Exchange 100 Index (the FTSE).

She went on to say that this new axis of power is intrinsically neither good nor bad: what matters is how it wields its power. It helped to raise China's standard of living, which many of us would consider to be beneficial. It also encouraged a worldwide abandonment of welfare societies, replacing them by a shareholder culture, which many of us would consider to be harmful. A less controversial example of a bad outcome is the Enron scandal, which broke in 2001. Enron was an energy company based in Texas, and its collapse led to what was then the biggest bankruptcy in American history, and a loss to shareholders of $11 billion. Enron was another warning, this time about deregulated accounting laws. Again, few heeded the warning.

Poovey did. She pointed to the contrast between the traditional financial system based on the production of real goods, and the emerging new one based on investment, currency trading, and 'complex wagers that future prices would rise or fall'. By 1995 this economy of virtual money had overtaken the real economy of manufacturing. The new axis of power was deliberately confusing real and virtual money: arbitrary figures in company accounts and actual cash or commodities. This trend, she argued, was leading to a culture in which the values of both goods and financial instruments were becoming wildly unstable, liable to explode or collapse at the click of a mouse.

The article illustrated these points using five common financial techniques and instruments, such as 'mark to market accounting', in which a company sets up a partnership with a subsidiary. The subsidiary buys a stake in the parent company's future profits; the money involved is then recorded as instant earnings by the parent company while the risk is relegated to the subsidiary's balance sheet. Enron used this technique when it changed its marketing strategy from selling energy to selling energy futures. The big problem with bringing forward potential future profits in this manner is that they cannot then be listed as profits next year. The answer is to repeat the manoeuvre. It's like trying to drive a car without brakes by pressing ever harder on the accelerator. The inevitable result is a crash.

Poovey's fifth example was derivatives, and it was the most important of them all because the sums of money involved were so gigantic. Her analysis largely reinforces what I've already said. Her main conclusion was: 'Futures and derivatives trading depends upon the belief that the stock market behaves in a statistically predictable way, in other words, that mathematical equations accurately describe the market.' But she noted that the evidence points in a totally different direction: somewhere between 75% and 90% of all futures traders lose money in any year.

Two types of derivative were particularly implicated in creating the toxic financial markets of the early twenty-first century: credit default swaps and collateralised debt obligations. A credit default swap is a form of insurance: pay your premium and you collect from an insurance company if someone defaults on a debt. But anyone could take out such insurance on anything. They didn't have to be the company that owed, or was owed, the debt. So a hedge fund could, in effect, bet that a bank's customers were going to default on their mortgage payments – and if they did, the hedge fund would make a bundle, even though it was not a party to the mortgage agreements. This provided an incentive for speculators to influence market conditions to make defaults more likely. A collateralised debt obligation is based on a collection (portfolio) of assets. These might be tangible, such as mortgages secured against real property, or they might be derivatives, or they might be a mixture of both. The owner of the assets sells investors the right to a share of the profits from those assets. The investor can play it safe, and get first call on the profits, but this costs them more. Or they can take a risk, pay less, and be lower down the pecking order for payment.

Both types of derivative were traded by banks, hedge funds, and other speculators. They were priced using descendants of the Black–Scholes equation, so they were considered to be assets in their own right. Banks borrowed money from other banks, so that they could lend it to people who wanted mortgages; they secured these loans with real property and fancy derivatives. Soon everyone was lending huge sums of money to everyone else, much of it secured on financial derivatives. Hedge funds and other speculators were trying to make money by spotting potential disasters and betting that they would happen. The value of the derivatives concerned, and of real assets such as property, was often calculated on a mark to market basis, which is open to abuse because it uses artificial accounting procedures and risky subsidiary companies to represent estimated future profit as actual present-day profit. Nearly everyone in the business assessed how risky the derivatives were using the same method, known as 'value at risk'. This calculates the probability that the investment might make a loss that exceeds some specified threshold. For example, investors might be willing to accept a loss of a million dollars if its probability were less than 5%, but not if it were more likely. Like Black–Scholes, value at risk assumes that there are no fat tails. Perhaps the worst feature was that the entire financial sector was estimating its risks using exactly the same method. If the method were at fault, this would create a shared delusion that the risk was low when in reality it was much higher.

It was a train crash waiting to happen, a cartoon character who had walked a mile off the edge of the cliff and remained suspended in mid-air only because he flatly refused to take a look at what was under his feet. As Poovey and others like her had repeatedly warned, the models used to value the financial products and estimate their risks incorporated simplifying assumptions that did not accurately represent real markets and the dangers inherent in them. Players in the financial markets ignored these warnings. Six years later, we all found out why this was a mistake.

Perhaps there is a better way.

The Black–Scholes equation changed the world by creating a booming quadrillion-dollar industry; its generalisations, used unintelligently by a small coterie of bankers, changed the world again by contributing to a multitrillion-dollar financial crash whose ever more malign effects, now extending to entire national economics, are still being felt worldwide. The equation belongs to the realm of classical continuum mathematics, having its roots in the partial differential equations of mathematical physics. This is a realm in which quantities are infinitely divisible, time flows continuously, and variables change smoothly. The technique works for mathematical physics, but it seems less appropriate to the world of finance, in which money comes in discrete packets, trades occur one at a time (albeit very fast), and many variables can jump erratically.

The Black–Scholes equation is also based on the traditional assumptions of classical mathematical economics: perfect information, perfect rationality, market equilibrium, the law of supply and demand. The subject has been taught for decades as if these things are axiomatic, and many trained economists have never questioned them. Yet they lack convincing empirical support. On the few occasions when anyone does experiments to observe how people make financial decisions, the classical scenarios usually fail. It's as though astronomers had spent the last hundred years calculating how planets move, based on what they thought was reasonable, without actually taking a look to see what they really did.

It's not that classical economics is completely wrong. But it's wrong more often that its proponents claim, and when it does go wrong, it goes very wrong indeed. So physicists, mathematicians, and economists are looking for better models. At the forefront of these efforts are models based on complexity science, a new branch of mathematics that replaces classical continuum thinking by an explicit collection of individual agents, interacting according to specified rules.

A classical model of the movement of the price of some commodity, for example, assumes that at any instant there is a single 'fair' price, which in principle is known to everyone, and that prospective purchasers compare this price with a utility function (how useful the commodity is to them) and buy it if its utility outweighs its cost. A complex system model is very different. It might involve, say, ten thousand agents, each with its own view of what the commodity is worth and how desirable it is. Some agents would know more than others, some would have more accurate information than others; many would belong to small networks that traded information (accurate or not) as well as money and goods.

A number of interesting features have emerged from such models. One is the role of the herd instinct. Market traders tend to copy other market traders. If they don't, and it turns out that the others are on to a good thing, their bosses will be unhappy. On the other hand, if they follow the herd and everyone's got it wrong, they have a good excuse: it's what everyone else was doing. Black–Scholes was perfect for the herd instinct. In fact, virtually every financial crisis in the last century has been pushed over the edge by the herd instinct. Instead of some banks investing in property and others in manufacturing, say, they *all* rush into property. This overloads the market, with too much money seeking too little property, and the whole thing comes to bits. So now they all rush into loans to Brazil, or to Russia, or back into a newly revived property market, or lose their collective marbles over dotcom companies – three kids in a room with a computer and a modem being valued at ten times the worth of a major manufacturer with a real product, real customers, and real factories and offices. When that goes belly-up, they *all* rush into the subprime mortgage market . . .

That's not hypothetical. Even as the repercussions of the global banking crisis reverberate through ordinary people's lives, and national economies flounder, there are signs that no lessons have been learned. A rerun of the dotcom fad is in progress, now aimed at social networking websites: Facebook has been valued at $100 billion, and Twitter (the website where celebrities send 140-character 'tweets' to their devoted followers) has been valued at $8 billion despite never having made a profit. The International Monetary Fund has also issued a strong warning about exchange traded funds (ETFs), a very successful way to invest in commodities like oil, gold, or wheat without actually buying any. All of these have gone up in price very rapidly, providing big profits for pension funds and other large investors, but the IMF has warned that these investment vehicles have 'all the hallmarks of a bubble waiting to burst . . . reminiscent of what happened in the securitisation market before the crisis'. ETFs are very like the derivatives that triggered the credit crunch, but secured in commodities rather than property. The stampede into ETFs has driven commodity prices through the roof, inflating them out of all proportion to the real demand. Many people in the third world are now unable to afford staple foodstuffs because speculators in developed countries are taking big gambles on wheat. The ousting of Hosni Mubarak in Egypt was to some extent triggered by huge increases in the price of bread.

The main danger is that ETFs are starting to be repackaged into further derivatives, like the collateralised debt obligations and credit default swaps that burst the subprime mortgage bubble. If the commodities bubble bursts, we could see a rerun of the collapse: just delete 'property' and insert 'commodities'. Commodity prices are very volatile, so ETFs are high-risk investments – not a great choice for a pension fund. So once again investors are being encouraged to take ever more complex, and ever more risky, bets, using money they don't have to buy stakes in things they don't want and can't use, in pursuit of speculative profits – while the people who do want those things can no longer afford them.

Remember the Dojima rice exchange?

Economics is not the only area to discover that its prized traditional theories no longer work in an increasingly complex world, where the old rules no longer apply. Another is ecology, the study of natural systems such as forests or coral reefs. In fact, economics and ecology are uncannily similar in many respects. Some of the resemblance is illusory: historically each has often used the other to justify its models, instead of comparing the models with the real world. But some is real: the interactions between large numbers of organisms are very like those between large numbers of stock market traders.

This resemblance can be used as an analogy, in which case it is dangerous because analogies often break down. Or it can be used as a source of inspiration, borrowing modelling techniques from ecology and applying them in suitably modified form to economics. In January 2011, in the journal *Nature*, Andrew Haldane and Robert May outlined some possibilities.[4] Their arguments reinforce several of the messages earlier in this chapter, and suggest ways of improving the stability of financial systems.

Haldane and May looked at an aspect of the financial crisis that I've not yet mentioned: how derivatives affect the stability of the financial system. They compare the prevailing view of orthodox economists, which maintain that the market automatically seeks a stable equilibrium, with a similar view in 1960s ecology, that the 'balance of nature' tends to keep ecosystems stable. Indeed, at that time many ecologists thought that any sufficiently complex ecosystem would be stable in this way, and that unstable behaviour, such as sustained oscillations, implied that the system was insufficiently complex. We saw in Chapter 16 that this is wrong. In fact, current understanding indicates exactly the opposite. Suppose that a large number of species interact in an ecosystem. As the network of ecological interactions becomes more complex through the addition of new links between species, or the interactions become stronger, there is a sharp threshold beyond which the ecosystem ceases to be stable. (Here chaos counts as stability; fluctuations can occur provided they remain within specific limits.) This discovery led ecologists to look for special types of interaction network, unusually conducive to stability.

Might it be possible to transfer these ecological discoveries to global finance? There are close analogies, with food or energy in an ecology corresponding to money in a financial system. Haldane and May were aware that this analogy should not be used directly, remarking: 'In financial ecosystems, evolutionary forces have often been survival of the fattest rather than the fittest.' They decided to construct financial models not by mimicking ecological models, but by exploiting the general modelling principles that had led to a better understanding of ecosystems.

They developed several economic models, showing in each case that under suitable circumstances, the economic system would become unstable. Ecologists deal with an unstable ecosystem by managing it in a way that creates stability. Epidemiologists do the same with a disease epidemic; this is why, for example, the British government developed a policy of controlling the 2001 foot-and-mouth epidemic by rapidly slaughtering cattle on farms near any that proved positive for the disease, and stopping all movement of cattle around the country. So government regulators' answer to an unstable financial system should be to take action to stabilise it. To some extent they are now doing this, after an initial panic in which they threw huge amounts of taxpayers' money at the banks but omitted to impose any conditions beyond vague promises, which have not been kept.

However, the new regulations largely fail to address the real problem, which is the poor design of the financial system itself. The facility to transfer billions at the click of a mouse may allow ever-quicker profits, but it also lets shocks propagate faster, and encourages increasing complexity. Both of these are destabilising. The failure to tax financial transactions allows traders to exploit this increased speed by making bigger bets on the market, at a faster rate. This also tends to create instability. Engineers know that the way to get a rapid response is to use an unstable system: stability by definition indicates an innate resistance to change, whereas a quick response requires the opposite. So the quest for ever greater profits has caused an ever more unstable financial system to evolve.

Building yet again on analogies with ecosystems, Haldane and May offer some examples of how stability might be enhanced. Some correspond to the regulators' own instincts, such as requiring banks to hold more capital, which buffers them against shocks. Others do not; an example is the suggestion that regulators should focus not on the risks associated with individual banks, but on those associated with the entire financial system. The complexity of the derivatives market could be reduced by requiring all transactions to pass through a centralised clearing agency. This would have to be extremely robust, supported by all major nations, but if it were, then propagating shocks would be damped down as they passed through it.

Another suggestion is increased diversity of trading methods and risk assessment. An ecological monoculture is unstable because any shock that occurs is likely to affect everything simultaneously, in the same way. When all banks are using the same methods to assess risk, the same problem arises: when they get it wrong, they all get it wrong at the same time. The financial crisis arose in part because all of the main banks were funding their potential liabilities in the same way, assessing the value of their assets in the same way, and assessing their likely risk in the same way.

The final suggestion is modularity. It is thought that ecosystems stabilise themselves by organising (through evolution) into more or less self-contained modules, connected to each other in a fairly simple manner. Modularity helps to prevent shocks propagating. This is why regulators worldwide are giving serious consideration to breaking up big banks and replacing them by a number of smaller ones. As Alan Greenspan, a distinguished American economist and former chairman of the Federal Reserve of the USA said of banks: 'If they're too big to fail, they're too big.'

Was an equation to blame for the financial crash, then?

An equation is a tool, and like any tool, it has to be wielded by someone how knows how to use it, and for the right purpose. The Black–Scholes equation may have contributed to the crash, but only because it was abused. It was no more responsible for the disaster than a trader's computer would have been if its use led to a catastrophic loss. The blame for the failure of tools should rest with those who are responsible for their use. There is a danger that the financial sector may turn its back on mathematical analysis, when what it actually needs is a better range of models, and – crucially – a solid understanding of their limitations. The financial system is too complex to be run on human hunches and vague reasoning. It desperately needs *more* mathematics, not less. But it also needs to learn how to use mathematics intelligently, rather than as some kind of magical talisman.

Where Next?

When someone writes down an equation, there isn't a sudden clap of thunder after which everything is different. Most equations have little or no effect (I write them down all the time, and believe me, I know). But even the greatest and most influential equations need help to change the world – efficient ways to solve them, people with the imagination and drive to exploit what they tell us, machinery, resources, materials, money. Bearing this in mind, equations have repeatedly opened up new directions for humanity, and acted as our guides as we explore them.

It took a lot more than seventeen equations to get us where we are today. My list is a selection of some of the most influential, and each of them required a host of others before it became seriously useful. But each of the seventeen fully deserves inclusion, because it played a pivotal role in history. Pythagoras led to practical methods for surveying our lands and navigating our way to new ones. Newton tells us how planets move and how to send space probes to explore them. Maxwell provided a vital clue that led to radio, TV, and modern communications. Shannon derived unavoidable limits to how efficient those communications can be.

Often, what an equation led to was quite different from what interested its inventor/discoverers. Who would have predicted in the fifteenth century that a baffling, apparently impossible number, stumbled upon while solving algebra problems, would be indelibly linked to the even more baffling and apparently impossible world of quantum physics – let alone that this would pave the road to miraculous devices that can solve a million algebra problems every second, and let us instantly be seen and heard by friends on the other side of the planet? How would Fourier have reacted if he had been told that his new method for studying heat flow would be built into machines the size of a pack of cards, able to paint extraordinarily accurate and detailed pictures of anything they are pointed at – in colour, even *moving*, with thousands of them contained in something the size of a coin?

Equations trigger events, and events, to paraphrase former British Prime Minister Harold Macmillan, are what keep us awake at night. When a revolutionary equation is unleashed, it develops a life of its own. The consequences can be good or bad, even when the original intention was benevolent, as it was for every one of my seventeen. Einstein's new physics gave us a new understanding of the world, but one of the things we used it for was nuclear weapons. Not as directly as popular myth claims, but it played its part nonetheless. The Black–Scholes equation created a vibrant financial sector and then threatened to destroy it. Equations are what we make of them, and the world can be changed for the worse as well as for the better.

Equations come in many kinds. Some are mathematical truths, tautologies: think of Napier's logarithms. But tautologies can still be powerful aids to human thought and deed. Some are statements about the physical world, which for all we know could have been different. Equations of this kind tell us nature's laws, and solving them tells us the consequences of those laws. Some have both elements: Pythagoras's equation is a theorem in Euclid's geometry, but it also governs measurements made by surveyors and navigators. Some are little better than definitions – but i and information tell us a great deal, once we have defined them.

Some equations are universally valid. Some describe the world very accurately, but not perfectly. Some are less accurate, confined to more limited realms, yet offer vital insights. Some are basically plain wrong, yet they can act as stepping-stones to something better. They may still have a huge effect.

Some even open up difficult questions, philosophical in nature, about the world we live in and our own place within it. The problem of quantum measurement, dramatised by Schrödinger's hapless cat, is one such. The second law of thermodynamics raises deep issues about disorder and the arrow of time. In both cases, some of the apparent paradoxes can be resolved, in part, by thinking less about the content of the equation and more about the context in which it applies. Not the symbols, but the boundary conditions. The arrow of time is not a problem about entropy: it's a problem about the context in which we *think* about entropy.

Existing equations can acquire new importance. The search for fusion power, as a clean alternative to nuclear power and fossil fuels, requires an understanding of how extremely hot gas, forming a plasma, moves in a magnetic field. The atoms of the gas lose electrons and become electrically charged. So the problem is one in magnetohydrodynamics, requiring a combination of the existing equations for fluid flow and for electromagnetism. The combination leads to new phenomena, suggesting how to keep the plasma stable at the temperatures needed to produce fusion. The equations are old favourites.

There is (or may be) one equation, above all, that physicists and cosmologists would give their eye teeth to lay hands on: a Theory of Everything, which in Einstein's day was called a Unified Field Theory. This is the long-sought equation that unifies quantum mechanics and relativity, and Einstein spent his later years in a fruitless quest to find it. These two theories are both successful, but their successes occur in different domains: the very small and the very large. When they overlap, they are incompatible. For example, quantum mechanics is linear, relativity isn't. Wanted: an equation that explains why both are so successful, but does the job of both with no logical inconsistencies. There are many candidates for a Theory of Everything, the best known being the theory of superstrings. This, among other things, introduces extra dimensions of space: six of them, seven in some versions. Superstrings are mathematically elegant, but there is no convincing evidence for them as a description of nature. In any case, it is desperately hard to carry out the calculations needed to extract quantitative predictions from superstring theory.

For all we know, there may not be a Theory of Everything. All of our equations for the physical world may just be oversimplified models, describing limited realms of nature in a way that we can understand, but not capturing the deep structure of reality. Even if nature truly obeys rigid laws, they might not be expressible as equations.

Even if equations are relevant, they need not be simple. They might be so complicated that we can't even write them down. The 3 billion DNA bases of the human genome are, in a sense, part of the equation for a human being. They are parameters that might be inserted into a more general equation for biological development. It is (barely) possible to print the genome on paper; it would need about two thousand books the size of this one. It fits into a computer memory fairly easily. But it's only one tiny part of any hypothetical human equation.

When equations become that complex, we need help. Computers are already extracting equations from big sets of data, in circumstances where the usual human methods fail or are too opaque to be useful. A new approach called evolutionary computing extracts significant patterns: specifically, formulas for conserved quantities – things that don't change. One such system called Eureqa, formulated by Michael Schmidt and Hod Lipson, has scored some successes. Software like this might help. Or it might not lead anywhere that really matters.

Some scientists, especially those with backgrounds in computing, think that it's time we abandoned traditional equations altogether, especially continuum ones like ordinary and partial differential equations. The future is discrete, it comes in whole numbers, and the equations should give way to algorithms – recipes for calculating things. Instead of solving the equations, we should simulate the world digitally by running the algorithms. Indeed, the world itself may *be* digital. Stephen Wolfram made a case for this view in his controversial book *A New Kind of Science*, which advocates a type of complex system called a cellular automaton. This is an array of cells, typically small squares, each existing in a variety of distinct states. The cells interact with their neighbours according to fixed rules. They look a bit like an eighties computer game, with coloured blocks chasing each other over the screen.

Wolfram puts forward several reasons why cellular automata should be superior to traditional mathematical equations. In particular, some of them can carry out any calculation that could be performed by a computer, the simplest being the famous Rule 110 automaton. This can find successive digits of π, solve the three-body equations numerically, implement the Black–Scholes formula for a call option – whatever. Traditional methods for solving equations are more limited. I don't find this argument terribly convincing, because it is also true that any cellular automaton can be simulated by a traditional dynamical system. What counts is not whether one mathematical system can simulate another, but which is most effective for solving problems or providing insights. It's quicker to sum a traditional series for π by hand than it is to calculate the same number of digits using the Rule 110 automaton.

However, it is still entirely credible that we might soon find new laws of nature based on discrete, digital structures and systems. The future may consist of algorithms, not equations. But until that day dawns, if ever, our greatest insights into nature's laws take the form of equations, and we should learn to understand them and appreciate them. Equations have a track record. They really have changed the world – and they will change it again.

Notes

Chapter 1

1 *The Penguin Book of Curious and Interesting Mathematics* by David Wells quotes a brief form of the joke: An Indian chief had three wives who were preparing to give birth, one on a buffalo hide, one on a bear hide, and the third on a hippopotamus hide. In due course, the first gave him a son, the second a daughter, and the third, twins, a boy and a girl, thereby illustrating the well-known theorem that the squaw on the hippopotamus is equal to the sum of the squaws on the other two hides. The joke goes back at least to the mid-1950s, when it was broadcast in the BBC radio series 'My Word', hosted by comedy scriptwriters Frank Muir and Denis Norden.

2 Quoted without reference on:
http://www-history.mcs.st-and.ac.uk/HistTopics/Babylonian_Pythagoras.html

3 A. Sachs, A. Goetze, and O. Neugebauer. *Mathematical Cuneiform Texts*, American Oriental Society, New Haven 1945.

4 The figure is repeated for convenience in Figure 60.

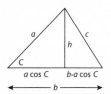

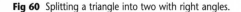

Fig 60 Splitting a triangle into two with right angles.

The perpendicular cuts the side *b* into two pieces. By trigonometry, one piece has length $a \cos C$, so the other has length $b\text{-}a \cos C$. Let *h* be the height of the perpendicular. By Pythagoras:

$$a^2 = h^2 + (a \cos C)^2$$
$$c^2 = h^2 + (b - a \cos C)^2$$

That is,

$$a^2 - h^2 = a^2 \cos^2 C$$
$$c^2 - h^2 = (b - a \cos C)^2 = b^2 - 2ab \cos C + a^2 \cos^2 C$$

Subtract the first equation from the second; now the unwanted h^2 cancels out. So do the terms $a^2 \cos^2 C$, and we are left with

$$c^2 - a^2 = b^2 - 2ab \cos C$$

which leads to the stated formula.

Chapter 2

1 http://www.17centurymaths.com/contents/napiercontents.html

2 Quoted from a letter John Marr wrote to William Lilly.

3 Prosthapheiresis was based on a trigonometric formula discovered by François Viète, namely

$$\sin\frac{x+y}{2} \cos\frac{x-y}{2} = \frac{\sin x + \sin y}{2}$$

If you owned a table of sines, the formula allowed you to calculate any product using only sums, differences, and division by 2.

Chapter 3

1 Keynes never delivered the lecture. The Royal Society planned to commemorate Isaac Newton's tercentenary in 1942, but World War II intervened, so the celebrations were postponed to 1946. The lecturers were the physicists Edward da Costa Andrade and Niels Bohr, and the mathematicians Herbert Turnbull and Jacques Hadamard. The society had also invited Keynes, whose interests included Newton's manuscripts as well as economics. He had written a lecture with the title 'Newton, the man', but he died just before the event took place. His brother Geoffrey read the lecture on his behalf.

2 This phrase comes from a letter that Newton wrote to Hooke in 1676. It wasn't new: in 1159 John of Salisbury wrote that 'Bernard of Chartres used to say that we are like dwarfs on the shoulders of giants, so that we can see more than they.' By the seventeenth century it had become a cliché.

3 Division by zero leads to fallacious proofs. For example, we can 'prove' that all numbers are zero. Assume that $a = b$. Therefore $a^2 = ab$, so $a^2 - b^2 = ab - b^2$. Factorise to get $(a+b)(a-b) = b(a-b)$. Divide by $(a-b)$ to deduce that $a+b = b$. Therefore $a = 0$. The error is the division by $(a-b)$, which is 0 because we assumed $a = b$.

4 Richard Westfall. *Never at Rest*, Cambridge University Press, Cambridge 1980, p. 425.

5 Erik H. Hauri, Thomas Weinreich, Alberto E. Saal, Malcolm C. Rutherford, and James A. Van Orman. High pre-eruptive water contents preserved in lunar melt

inclusions, *Science Online* (26 May 2011) 1204626. [DOI:10.1126/science.1204626]. Their results proved controversial.

6   However, it's not coincidence. It works for any differentiable function: one with a continuous derivative. These include all polynomials and all convergent power series, such as the logarithm, the exponential, and the various trigonometric functions.

7   The modern definition is: a function $f(h)$ tends to a limit L as h tends to zero if for any $\varepsilon > 0$ there exists $\delta > 0$ such that $|h| < \delta$ implies that $|f(h) - L| < \varepsilon$. Using *any* $\varepsilon > 0$ avoids referring to anything flowing or becoming smaller: it deals with all possible values in one go.

Chapter 4

1   The book of Genesis refers to the 'firmament'. Most scholars think this derives from the ancient Hebrew belief that the stars were tiny lights fixed to a solid vault of Heaven, shaped like a hemisphere. This is what the night sky looks like: the way our visual senses respond to distant objects makes the stars appear to be at much the same distance from us. Many cultures, especially in the Middle and Far East, thought of the heavens as a slowly spinning bowl.

2   The Great Comet of 1577 is not Halley's comet, but another of historical importance, now called C/1577 V1. It was visible to the naked eye in 1577 AD. Brahe observed the comet and deduced that comets were located outside the Earth's atmosphere. The comet is currently about 24 billion kilometres from the Sun.

3   The figure was not known until 1798, when Henry Cavendish obtained a reasonably accurate value in a laboratory experiment. It is about 6.67×10^{-11} newton metre squared per kilogram squared.

4   June Barrow-Green. *Poincaré and the Three Body Problem*, American Mathematical Society, Providence 1997.

Chapter 5

1   In 1535 the mathematicians Antonio Fior and Niccolò Fontana (nicknamed Tartaglia, 'the stammerer') engaged in a public contest. They set each other cubic equations to solve, and Tartaglia beat Fior comprehensively. At that time, cubic equations were classified into three distinct types, because negative numbers were not recognised. Fior knew how to solve just one type; initially Tartaglia knew how to solve one different type, but shortly before the contest he figured out how to solve all the other types. He then set Fior only the types that he knew Fior could not solve. Cardano, working on his algebra text, heard about the contest, and realised that Fior and Tartaglia knew how to solve cubics. This discovery would greatly enhance the book, so he asked Tartaglia to reveal his methods.

Eventually Tartaglia divulged the secret, later stating that Cardano had promised never to make it public. But the method appeared in the *Ars Magna*, so Tartaglia accused Cardano of plagiarism. However, Cardano had an excuse, and he also had a good reason to find a way round his promise.

His student Lodovico Ferrari had found how to solve quartic equations, an equally novel and dramatic discovery, and Cardano wanted that in his book, too. However, Ferrari's method required the solution of an associated cubic equation, so Cardano could not publish Ferrari's work without also publishing Tartaglia's.

Then he learned that Fior was a student of Scipio del Ferro, who was rumoured to have solved all three types of cubic, passing just one type on to Fior. Del Ferro's unpublished papers were in the possession of Annibale del Nave. So Cardano and Ferrari went to Bologna in 1543 to consult del Nave, and in the papers they found solutions to all three types of cubic. So Cardano could honestly say that he was publishing del Ferro's method, not Tartaglia's. Tartaglia still felt cheated, and published a long, bitter diatribe against Cardano. Ferrari challenged him to a public debate and won hands down. Tartaglia never really recovered his reputation after that.

Chapter 6

1   Summarised in Chapter 12 of: Ian Stewart. *Mathematics of Life*, Profile, London 2011.

Chapter 7

1   Yes, I know this is the plural of 'die', but nowadays everyone uses it for the singular as well, and I've given up fighting this tendency. It could be worse: someone just sent me an e-mail carefully using 'dice' for the singular and 'die' for the plural.

2   There are many fallacies in Pascal's argument. The main one is that it would apply to any hypothetical supernatural being.

3   The theorem states that under certain (fairly common) conditions, the sum of a large number of random variables will have an approximately normal distribution. More precisely, if $(x_1, \ldots, x_n)$ is a sequence of independent identically distributed random variables, each having mean μ and variance σ^2, then the central limit theorem states that

$$\sqrt{n}\left(\frac{1}{n}\sum_{i=1}^{n} x_i - \mu\right)$$

converges to a normal distribution with mean 0 and standard deviation σ as n becomes arbitrarily large.

Chapter 8

1   Look at three consecutive masses, numbered $n-1$, n, $n+1$. Suppose that at time t they are displaced distances $u_{n-1}(t)$, $u_n(t)$, and $u_{n+1}(t)$ from their initial positions on the horizontal axis. By Newton's second law the acceleration of each mass is proportional to the forces that act on it. Make the simplifying assumption that each mass moves through a very small distance in the vertical direction only. To a very good approximation, the force that mass $n-1$ exerts on mass n is then proportional to the difference $u_{n-1}(t) - u_n(t)$, and similarly the

force that mass $n+1$ exerts on mass n is proportional to the difference $u_{n+1}(t) - u_n(t)$. Adding these together, the total force exerted on mass n is proportional to $u_{n-1}(t) - 2u_n(t) + u_{n+1}(t)$. This is the difference between $u_{n-1}(t) - u_n(t)$ and $u_n(t) - u_{n+1}(t)$, and each of these expressions is also the difference between the positions of consecutive masses. So the force exerted on mass n is a *difference between differences*.

Now suppose the masses are very close together. In calculus, a difference – divided by a suitable small constant – is an approximation to a derivative. A difference between differences is an approximation to a derivative of a derivative, that is, a second derivative. In the limit of infinitely many point masses, infinitesimally close together, the force exerted at a given point of the spring is therefore proportional to $\partial^2 u/\partial x^2$, where x is the space coordinate measured along the length of the string. By Newton's second law this is proportional to the acceleration at right angles to that line, which is the second time derivative $\partial^2 u/\partial t^2$. Writing the constant of proportionality as c^2 we get

$$\frac{\partial^2 u}{\partial t^2} = c^2 \frac{\partial^2 u}{\partial x^2}$$

where $u(x,t)$ is the vertical position of location x on the string at time t.

2   For an animation see http://en.wikipedia.org/wiki/Wave_equation

3   In symbols, the solutions are precisely the expressions

$$u(x,t) = f(x - ct) + g(x + ct)$$

for any functions f and g.

4   Animations of the first few normal modes of a circular drum can be found at http://en.wikipedia.org/wiki/Vibrations_of_a_circular_drum

Circular and rectangular drum animations are at http://www.mobiusilearn.com/viewcasestudies.aspx?id=2432

Chapter 9

1   Suppose that $u(x,t) = e^{-n^2 \alpha t} \sin nx$. Then

$$\frac{\partial u}{\partial t} = -n^2 \alpha e^{-n^2 \alpha t} \sin nx = \frac{\partial^2 u}{\partial x^2}$$

Therefore $u(x,t)$ satisfies the heat equation.

2   This is JFIF encoding, used for the web. EXIF coding, for cameras, also includes 'metadata' describing the camera settings, such as date, time, and exposure.

Chapter 10

1   http://www.nasa.gov/topics/earth/features/2010-warmest-year.html

Chapter 11

1   Donald McDonald. How does a cat fall on its feet?, *New Scientist* **7** no. 189 (1960) 1647–9. See also http://en.wikipedia.org/wiki/Cat_righting_reflex

2   The curl of both sides of the third equation gives

$$\nabla \times \nabla \times \mathbf{E} = -\frac{1}{c}\frac{\partial(\nabla \times \mathbf{H})}{\partial t}$$

Vector calculus tells us that the left-hand side of this equation simplifies to

$$\nabla \times \nabla \times \mathbf{E} = \nabla(\nabla \cdot \mathbf{E}) - \nabla^2\mathbf{E} = -\nabla^2\mathbf{E}$$

where we also use the first equation. Here ∇^2 is the Laplace operator. Using the fourth equation, the right-hand side becomes

$$-\frac{1}{c}\frac{\partial(\nabla \times \mathbf{H})}{\partial t} = -\frac{1}{c}\frac{\partial}{\partial t}\left(\frac{1}{c}\frac{\partial \mathbf{E}}{\partial t}\right) = -\frac{1}{c^2}\frac{\partial^2 \mathbf{E}}{\partial t^2}$$

Cancelling out two minus signs and multiplying by c^2 yields the wave equation for **E**:

$$\frac{\partial^2 \mathbf{E}}{\partial t^2} = c^2\nabla^2\mathbf{E}$$

A similar calculation yields the wave equation for **H**.

Chapter 12

1   Specifically,

$$S_A - S_B = \int_A^B \frac{dq}{T}$$

where S_A and S_B are the entropies in states A and B.

2   The second law of thermodynamics is technically an *inequality*, not an equation. I've included the second law in this book because its central position in science demanded its inclusion. It is undeniably a mathematical *formula*, a loose interpretation of 'equation' that is widespread outside the technical scientific literature. The formula alluded to in Note 1 of this chapter, using an integral, is a genuine equation. It defines the change in entropy, but the second law tells us what its most important feature is.

3   Brown was anticipated by the Dutch physiologist Jan Ingenhousz, who saw the same phenomenon in coal dust floating on the surface of alcohol, but he didn't propose any theory to explain what he had seen.

Chapter 13

1   In the Gran Sasso National Laboratory, in Italy, is a 1300-tonne particle detector called OPERA (oscillation project with emulsion-tracking apparatus). Over two years it tracked 16,000 neutrinos produced at CERN, the European particle physics laboratory in Geneva. Neutrinos are electrically neutral

subatomic particles with a very small mass, and they can pass through ordinary matter with ease. The results were baffling: on average the neutrinos completed the 730-kilometre trip 60 nanoseconds (billionths of a second) sooner than they would have done if they had been travelling at the speed of light. The measurements are accurate to within 10 nanoseconds, but there remains the possibility of some systematic error in the way the times are calculated and interpreted, which is highly complex.

The results have been posted online: 'Measurement of the neutrino velocity with the OPERA detector in the CNGS beam' by the OPERA Collaboration, http://arxiv.org/abs/1109.4897

This article does not claim to have disproved relativity: it merely presents its observations as something that the team cannot explain with conventional physics. A non-technical report can be found at http://www.nature.com/news/2011/110922/full/news.2011.554.html

A possible source of systematic error, related to differences in the force of gravity at the two laboratories, is proposed at http://www.nature.com/news/2011/111005/full/news.2011.575.html but the OPERA team disputes this suggestion.

Most physicists think that, despite the great care exercised by the researchers, some systematic error is involved. In particular, previous observations of neutrinos from a supernova seem to conflict with the new ones. The resolution of the controversy will require independent experiments, and these will take several years. Theoretical physicists are already analysing potential explanations ranging from minor, well-known extensions of the standard model of particle physics to exotic new physics in which the universe has more dimensions than the usual four. By the time you read this, the story will already have moved on.

2   A thorough explanation is given by Terence Tao on his website: http://terrytao.wordpress.com/2007/12/28/einsteins-derivation-of-emc2/

The derivation of the equation involves five steps:
(a) Describe how space and time coordinates transform when the frame of reference is changed.
(b) Use this description to work out how the frequency of a photon transforms when the frame of reference is changed.
(c) Use Planck's law to work out how the energy and momentum of a photon transform.
(d) Apply conservation of energy and momentum to work out how the energy and momentum of a moving body transform.
(e) Fix the value of an otherwise arbitrary constant in the calculation by comparing the results with Newtonian physics when the velocity of the body is small.

3   Ian Stewart and Jack Cohen. *Figments of Reality*, Cambridge University Press, Cambridge 1997, page 37.

4   http://en.wikipedia.org/wiki/Mass%E2%80%93energy_equivalence

5   A few didn't see it that way. Henry Courten, reanalysing photographs of the 1970 solar eclipse, reported the existence of at least seven very tiny bodies in

close orbits round the Sun – perhaps evidence of a thinly populated inner asteroid belt. No conclusive evidence of their existence has been found, and they would have to be less than 60 kilometres across. The objects seen in the photographs may just have been passing small comets or asteroids in eccentric orbits. Whatever they were, they weren't Vulcan.

6 The vacuum energy in a cubic centimetre of free space is estimated to be 10^{-15} joules. According to quantum electrodynamics it should in theory be 10^{107} joules – wrong by a factor of 10^{122}.
http://en.wikipedia.org/wiki/Vacuum_energy

7 Penrose's work is reported in: Paul Davies. *The Mind of God*, Simon & Schuster, New York 1992.

8 Joel Smoller and Blake Temple. A one parameter family of expanding wave solutions of the Einstein equations that induces an anomalous acceleration into the standard model of cosmology. http://arxiv.org/abs/0901.1639

9 R.S. MacKay and C.P. Rourke. A new paradigm for the universe, preprint, University of Warwick 2011. For more details see the papers listed on http://msp.warwick.ac.uk/~cpr/paradigm/

Chapter 14

1 The Copenhagen interpretation is usually said to have emerged from discussions between Niels Bohr, Werner Heisenberg, Max Born, and others, in the mid-1920s. It acquired the name because Bohr was Danish, but none of the physicists involved used the term at the time. Don Howard has suggested that the name, and the viewpoint that it encapsulates, first appeared in the 1950s, probably through Heisenberg. See: D. Howard. 'Who Invented the "Copenhagen Interpretation"? A Study in Mythology', *Philosophy of Science* **71** (2004) 669–682.

2 Our cat Harlequin can often be observed in a superposition of the states 'asleep' and 'snoring', but that probably doesn't count.

3 Two science fiction novels about this are Philip K. Dick's *The Man in the High Castle* and Norman Spinrad's *The Iron Dream*. Thriller writer Len Deighton's *SS-GB* is also set in a counterfactual Nazi-ruled England.

Chapter 15

1 Suppose I roll a dice [see Note 1 of Chapter 7] and assign symbols a, b, c like this:

    a The dice rolls 1, 2, or 3
    b The dice rolls 4 or 5
    c The dice rolls 6

Symbol a occurs with probability $\frac{1}{2}$, symbol b has probability $\frac{1}{3}$, and symbol c has probability $\frac{1}{6}$. Then my formula, whatever it is, will assign an information content $H(\frac{1}{2}, \frac{1}{3}, \frac{1}{6})$.

However, I could think of this experiment in a different way. First I decide whether the dice rolls something less than or equal to 3, or greater. Call these possibilities q and r, so that

q   The dice rolls 1, 2, or 3
r   The dice rolls 4, 5, or 6

Now q has probability $\frac{1}{2}$ and r has probability $\frac{1}{2}$. Each conveys information H $(\frac{1}{2},\frac{1}{2})$. Case q is my original a, and case r is my original b and c. I can split case r into b and c, and their probabilities are $\frac{2}{3}$ and $\frac{1}{3}$ *given that r has happened.* If we now consider only this case, the information conveyed by whichever of b and c turns up is $H(\frac{2}{3}, \frac{1}{3})$. Shannon now insists that the original information should be related to the information in these subcases like this:

$$H\left(\tfrac{1}{2},\tfrac{1}{3},\tfrac{1}{6}\right) = H\left(\tfrac{1}{2},\tfrac{1}{2}\right) + \tfrac{1}{2}H\left(\tfrac{1}{2},\tfrac{1}{3}\right)$$

See Figure 61.

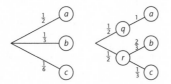

Fig 61 Combining choices in different ways. The information should be the same in either case.

The factor $\frac{1}{2}$ in front of the final H is present because this second choice occurs only half the time, namely when r is chosen in the first stage. There is no such factor in front of the H just after the equals sign, because this refers to a choice that is always made – between q and r.

2   See Chapter 2 of: C.E. Shannon and W. Weaver. *The Mathematical Theory of Communication*, University of Illinois Press, Urbana 1964.

Chapter 16

1   If the population x_t is relatively small, so that is close to zero, then $1 - x_t$ is close to 1. The next generation will therefore have a size close to kx_t, which is k times as large as the current one. As the size of the population increases, the extra factor $1 - x_t$ makes the actual growth rate smaller, and it drops to zero as the population approaches its theoretical maximum.

2   R.F. Costantino, R.A. Desharnais, J.M. Cushing, and B. Dennis. Chaotic dynamics in an insect population, *Science* **275** (1997) 389–391.

3   J. Huisman and F.J. Weissing. Biodiversity of plankton by species oscillations and chaos, *Nature* **402** (1999) 407–410.

4   E. Benincà, J. Huisman, R. Heerkloss, K.D. Jöhnk, P. Branco, E.H. Van Nes, M. Scheffer, and S.P. Ellner. Chaos in a long-term experiment with a plankton community, *Nature* **451** (2008) 822–825.

Chapter 17

1   The value of a call option is

$$C(s,t) = N(d_1)S - N(d_2)Ke^{-r(T-t)}$$

where

$$d_1 = \frac{\log(S/K) + (r + \sigma^2/2(T-t))}{\sigma\sqrt{T} - t}$$

$$d_2 = \frac{\log(S/K) + (r - \sigma^2/2(T-t))}{\sigma\sqrt{T} - t}$$

The price of a corresponding put option is

$$P(s,t) = [N(d_1) - 1]S + [1 - N(d_2)]Ke^{-r(T-t)}$$

Here $N(d_j)$ is the cumulative distribution function of the standard normal distribution for $j = 1, 2$, and $T-t$ is the time to maturity.

2   Strictly, a Sveriges Riksbank Prize in Economic Sciences in Memory of Alfred Nobel.

3   M. Poovey. Can numbers ensure honesty? Unrealistic expectations and the U.S. accounting scandal, *Notices of the American Mathematical Society* **50** (2003) 27–35.

4   A.G. Haldane and R.M. May. Systemic risk in banking ecosystems, *Nature* **469** (2011) 351–355.

Illustration Credits

Index

The suffix 'n' indicates a note. Italic page references refer to Figures.